Autonomy and Food Biotechnology in Theological Ethics

Cathriona Russell

Autonomy and Food Biotechnology in Theological Ethics

PETER LANG
Oxford • Bern • Berlin • Bruxelles • Frankfurt am Main • New York • Wien

Bibliographic information published by Die Deutsche Bibliothek
Die Deutsche Bibliothek lists this publication in the Deutsche Nationalbibliografie;
detailed bibliographic data is available on the Internet at <http://dnb.ddb.de>.

A catalogue record for this book is available from The British Library.

Library of Congress Cataloging-in-Publication Data:

Russell, Cathriona, 1965-
  Autonomy and food biotechnology in theological ethics / Cathriona
Russell.
    p. cm.
  Includes bibliographical references and index.
  ISBN 978-3-03911-838-0 (alk. paper)
  1. Human ecology--Religious aspects--Christianity. 2. Autonomy
(Psychology)--Religious aspects--Christianity. 3. Food--Biotechnology.
4. Genetically modified foods. I. Title.
  BT695.5.R86 2009
  241'.691--dc22
                                    2009007537

ISBN 978-3-03911-838-0

© Peter Lang AG, International Academic Publishers, Bern 2009
Hochfeldstrasse 32, Postfach 746, CH-3000 Bern 9, Switzerland
info@peterlang.com, www.peterlang.com, www.peterlang.net

All rights reserved.
All parts of this publication are protected by copyright.
Any utilisation outside the strict limits of the copyright law, without the permission of
the publisher, is forbidden and liable to prosecution.
This applies in particular to reproductions, translations, microfilming, and storage and
processing in electronic retrieval systems.

Printed in Germany

Contents

Acknowledgements 7

Introduction 9

1 Transgenics in Science and Economics 17
   Developments in transgenics 19
   Strategies in environmental management 43
   Sustainability: normative and descriptive aspects 58
   Conclusions 62

2 An Autonomy Perspective in Theological Ethics 65
   Divine command ethics or evangelical ethics 70
   Christian communitarian or ecclesial ethics 87
   Natural law 104
   Autonomy in Christian ethics 124
   Summary 143

3 Environmental Theologies and Reconstructions of Stewardship 149
   Biblical theology and the non-human creation 152
   Models for environmental ethics 161
   Reconstructions of 'stewardship' in theology 181
   Summary and Conclusions 205

4	Nature in Theological Perspective	209
	Nature as creation in the systematic theology of Wolfhart Pannenberg	213
	Nature as creation in the 'natural law' ethic of Michael Northcott	232
	Nature as creation in the 'virtue' ethic of Celia Deane-Drummond	251
	Towards a Christian anthropology of stewardship	263
	Conclusion	267
	Bibliography	273
	Index	287

Acknowledgements

My sincere thanks to Professor Maureen Junker-Kenny for her constant collegiality and to Professor Desmond O'Neill for his insights and enthusiasm. Thanks to Dr Vincent MacNamara and Dr John Scally for their helpful feedback on this work and to all the participants of the Graduate Seminar at the School of Religions and Theology, Trinity College Dublin.

This study was funded for three years by the Irish Research Council for the Humanities and Social Sciences and by a Trinity Foundation Scholarship. It was subsequently financed by a Broad Curriculum Studentship at TCD.

Introduction

This book outlines and defends the autonomy approach in Christian ethics and applies the insights of that approach in the ethically controversial area of the biotechnological manipulation of food organisms. Transgenic food organisms or genetically modified food (GMOs) are micro-organisms, plants and animals that have been transformed genetically using gene-transfer techniques. These recently developed techniques (collectively GM) are most commonly used to transfer genetic material (DNA) across classified 'species' barriers. They have revolutionised applied biology and they have also opened up avenues of investigation in basic biological research, for example in genome sequencing and in paleobiology. Modified micro-organisms have been used for several decades to manufacture human insulin, while insect-resistant crops, like cotton and corn, are also already a reality. Induced metal tolerance and drought resistance in plants and the improvement of the nutritional status of staple crops – as in the much publicised case of vitamin A-enriched rice – are in the offing. More ambitious long-term projects aim to induce the novel production of industrial polymers, pharmaceuticals and vaccines in plants.

This technology raises ethical issues in two related spheres: the biological and the societal. It has the potential to impact directly on human health, through the food chain, and indirectly on human and environmental health, through the potentially irreversible release of transgenic food organisms into the agricultural and wider environment. Along with the biological impact of transgenics, there are broader economic, social and legal issues at stake. Transgenic technology impacts on questions of consumer rights and the democratic regulation of technology, on economics and international trade, and on distributive justice in the developing world. It also has legal implications in its impact on intellectual property law in the international patenting system, although this theme would require a study of its own. The salient biological, regulatory and economic aspects of these novel technologies will be specifically taken up in this study and investigated from the perspective of an autonomy approach in Christian ethics.

In chapter one I will examine, from the perspective of the natural sciences and economics, the discoveries, benefits and unintended outcomes from the production of transgenic food organisms. The food safety and environmental management strategies of policy makers are critiqued and their potential to ameliorate the negative consequences accruing from these organisms are evaluated. This analysis illustrates that scientific and economic insights need to be fully integrated with other social and ethical imperatives. The concept of 'sustainability' will be discussed as a 'concept of convergence' for the natural sciences and for ethics. The task for the natural sciences is to contribute to the descriptive aspect of sustainability, which, it is argued, then needs to be ethically integrated with the normative aspects of sustainability in social and political policy formation.

In order to situate scientific and economic insights morally, it is necessary to argue for a position in ethics that has the capacity to integrate these disciplines, along with insights from theological and philosophical ethics. In chapter two I present resources in Christian ethics and philosophy that can contribute to such a framework, namely divine command morality, ecclesial communitarianism, natural law and autonomy. I specifically focus on the work of theologians who have taken up the theme of the moral significance of the non-human creation, in the context of the contemporary consciousness of environmental problems and the human instrumentalisation of the non-human creation.

I argue for the 'autonomy position in Christian ethics' as the most appropriate approach for the task of integrating the concerns articulated in these three positions. Undoubtedly there is more than one interpretation of autonomy in moral philosophy and ethics. The autonomy position that I outline has its roots in natural law ethics and in that sense it can be considered a continuation of that tradition in the context of the consciousness of freedom in modernity.[1] However, in contrast to natural law, the Christian autonomy position developed in this study is also indebted to Kantian moral philosophy. It takes Kant's analysis and critique of reason and his concept of autonomy as the philosophical counterpart for a theological understanding of the freedom of the human person before God. Kant's concept of autonomy and his categorical imperative are taken here as the most adequate philosophical expression of the moral significance of the dignity of the human person defended in Christian theology.

Kant's moral philosophy is based on a concept of the dignity of the person as moral self-legislator, but it also points beyond morality to the question of meaning and postulates, at the limits of reason, God, freedom and immortality. Theologically, the autonomy approach in Christian ethics places the emphasis on the human person's moral capacity as the receiver of God's self-revelation and as a creature destined for, not just liberation, but salvation in divine fellowship. Central to this autonomy approach is a concept of individual freedom that is defined by – not just limited by – the freedom of the other, and which can only be realised in solidarity. It cannot be read, therefore, as a type of atomistic individualism. It understands the human person as an 'end in themselves' in a 'kingdom of ends'. The autonomy position also recognises the positive role of human reason in morality. At the same time it shows the limits of reason and leaves room for an understanding of our moral autonomy before God, compatible with Christian ethics.

As a Christian ethics it shares the insights of other positions in theological ethics. It shares with Divine Command morality the idea that scriptural revelation and theological imagination shape Christian ethics, although it conceives of that in a specific way, as we shall see. In its

---

[1] The autonomy position can claim Thomas Aquinas therefore as a source. Proponents have argued, indeed, that the position of autonomy is not original but traditional, since like Aquinas it does not acknowledge any moral precepts that could not be known by reason. It can be considered therefore as a legitimate continuation of the tradition in modernity. Cf. Vincent MacNamara, *Faith and Ethics: Recent Roman Catholicism* (Dublin: Gill and MacMillan, 1985), 40.

theological reconstruction it recognises the role of participatory communities in moral and ethical formation, a theme that is central to communitarian critiques of individualism. In this regard it seeks a reintegration of an 'ethic of the good life' alongside questions of the right, of duty and obligation. The autonomy approach also values reason, but it suggests that it is not rationality as such but our moral autonomy that is the best philosophical expression of human dignity. The autonomy approach, therefore, can be distinguished from the other positions presented, in that it prioritises as its starting point, not scripture, or community, or 'nature as reason', but the autonomy and dignity of the human person. It can give a clear account of the interaction between the different constituents in moral and ethical discourse. It specifically argues that perspectives in theological ethics need to make explicit the philosophical and theological justifications, resources and themes they draw upon. It also happily acknowledges the debt that theology and philosophy owe each other in mutual elaboration and critique.

It may seem that a Kantian autonomy position would appear, at first glance, to be at odds with the theonomy implicit in Christian ethics and morality. Certainly the autonomy approach has been taken to task for arguing that Christian faith does not bring any specific content to morality. The question of a specific Christian content for morality must therefore be addressed. The autonomy position presented in this study argues that it is the transformative 'context' of faith that makes Christian morality distinctive in five dimensions. It provides heuristic potential, it motivates and sensitises us to put certain questions to ethics. It also helps to integrate and to relativise morality, as we shall see.

A Kantian autonomy approach might also seem incompatible with an 'environmentally friendly' view of non-human reality. For Kant it was the good will and not nature that was morally relevant. However, in its reintegration of the right and the good, the autonomy approach in its second generation presented here, recognises the place of non-human reality as the true context of finite human freedom. The concept of finitude is taken in this study to express both what is meant by 'embodiment' and the idea that nature is the true context of life. The autonomy approach therefore can encompass a view of the human person as 'embodied reason' in common with natural law approaches, but it gives us an anthropology of 'finite freedom' in contrast. This study therefore proposes, that the theological concept of 'creatureliness', be translated in ethics and philosophy as 'finite freedom'. This designation has the capacity to register the divine 'givenness' of our worldly existence but also the possibility of its transformation, a theme that is central to Christian creation theology. This will have important implications for environmental theology and the debate on transgenic food in the public realm.

In order to develop an ethic of the environment and of human production in a Christian context, it is necessary to examine approaches in theology that have articulated concern for the natural environment in light of its unprecedented destruction and degradation. Chapter three investigates some recent biblical approaches to environmental theology that can inform a Christian environmental ethic. In particular it examines the work of the biblical scholar Sean Freyne, who undertakes a reconstruction of the role that the natural world, specifically the non-human creation, played in the life and ministry of Jesus. Freyne's work highlights the centrality of the creation narratives in Jesus' self-understanding of his own scriptures and the theological imagination that could read the natural world he encountered as the gift of a benevolent creator, even in the face of hardship and poverty. This chapter also examines and critiques three typological approaches to human relationships with 'nature', anthropocentrism, ecocentrism and theocentrism. The potential of each to adequately capture Christian understandings of the human and non-human creation is assessed. It is argued that the Christian tradition is, and should remain, firmly anthropocentric. Nevertheless an emphasis on the place and special responsibility of the human person as the steward of creation can take on board the critiques both ecocentrism and theocentrism provide, where they strive to develop a non-instrumentalist view of the human natural context. The Christian environmental ethic I present is a qualified anthropocentrism, that values the human person as the steward of creation and where creation has the status of both a gift and a task. From an autonomy perspective, the appropriate interpretation of stewardship is neither domination nor co-creation, but service. It is precisely because it maintains the tensions and the dualism between the human person and the non-human creation, that the autonomy approach cannot lose sight of our responsibility for the non-human creation. In light of this, I will investigate how stewardship contributes to the normative aspect of sustainability (articulated in terms of a commitment to justice) and how secular intellectual resources can be positively appropriated in modern theology.

There is a growing consensus that agriculture needs to intensify in light of several conditions not least the world's increasing population, the understanding that the most productive land is already under cultivation and that with increasing standards of living there is increased demand for many foods such as meat, dairy, fruit and vegetables and that production of these may well be in competition with renewable biofuels. There is an assumption that

> ... additional high-yield practices, based on advances in biology, ecology, chemistry, and technology are critically needed in agriculture and forestry not only to achieve the goal of improving the human condition for all peoples but also the simultaneous preservation of the natural environment and its biodiversity through the conservation of wild areas and natural habitat.[2]

Yet improving the human condition for all (development) and environmental preservation (sustainability) have not always been viewed as compatible goals. I will examine under what conditions they are thought to be compatible and how their compatibility might influence the adoption of GM technology. Finally, I investigate the ethical implications of introducing transgenic crops to Irish agriculture in light of the commitment to sustainable development that is found to be in keeping with a Christian concept of stewardship.

Having discussed the role of the human person as steward, I will then investigate how environmental theologies have appropriated systematic

---

2   *The Declaration in Support of Protecting Nature with High-Yield Farming and Forestry* signed may 2002 by Dr N. Borlaug, Nobel peace laureate, Dr P. Moore, co-founder of Greenpeace, Mr E. Lapointe, World Conservation Trust, Senator G. McGovern, ambassador to the UN-FAO and Dr J. Lovelock, author of the Gaia theory. Available from www.greenspirit.com.

conceptualisations of nature 'as creation'. Theological critiques often focus on betrayals of stewardship. In chapter four I will treat the systematic foundations for a theological view of nature as creation, focusing in particular on the creation theology of Wolfhart Pannenberg. The objective is to examine how theologians have drawn on systematic themes to develop theological arguments in environmental ethics. An emphasis in environmental theology, for example, on such concepts as the 'prior order in nature' or the 'deep relationality' in nature does not do justice to the coherence and cogency of systematic conceptualisations of creation. There is a danger that other equally significant aspects of Christian creation theology, faith in a God who has created everything and 'who makes everything new', and the independence as well as interdependence in the 'world of creatures' will be neglected.

Pannenberg's creation theology is a worthy resource for this study on several grounds. It is written from the perspective of the modern consciousness of freedom and is historically and philosophically informed.[3] Pannenberg has also had a sustained and fruitful dialogue with the natural sciences, as an express part of his project. His insights into the relationship between science and theology in modernity are both integrated into his systematic theology and are the subject of specific works, for example in cosmology and in dialogue with scientific world views.[4] Pannenberg is also considered, from an ecumenical perspective, to have made an important contribution to a common Christian doctrine of creation.[5]

His creation theology is a reminder that the Christian tradition emphasises creation as a free act of God, who conserves but also transforms the original order. His approach also makes clear, that in the Christian

---

[3] Joseph Colombo, 'Systematic Theology' *The Journal of Religion* 76, January (1996), 125 and Donald Bloesch, 'A Review of Pannenberg's *Systematic Theology, Volume II*' *Christianity Today* 39, April (1995), 106.

[4] Cf. Wolfhart Pannenberg, *Towards a Theology of Nature* (Westminster: John Knox Press, 1993) and Wolfhart Pannenberg, 'Problems between Science and Theology in the Course of Their Modern History' *Zygon* 41, no. 1 (2006).

[5] Alexandre Ganoczy, 'Creation' in *Handbook of Catholic Theology*, ed. Wolfgang Beinert and Francis Schüssler Fiorenza (New York: Crossroads, 1995), 135.

understanding, the basic relationship of the human person to God is one of creatureliness both in our independence and our interdependence. This has significant implications for a theologically informed environmental ethics. The theological tradition does not simply call us to maintaining the 'prior order' and 'relationality' in nature, as some environmental theologies are tempted to propose. Against this systematic background I will test how successfully two environmental theologians, Michael Northcott and Celia Deane-Drummond, have interpreted and appropriated aspects of creation theology in developing their Christian environmental ethics. Northcott specifically builds on the divine command approach of Oliver O'Donovan in his analysis while Deane-Drummond builds a virtue ethics approach that is indebted to some communitarian and natural law themes.

The theological understanding of nature as creation has the potential to conceptualise and corroborate the concern for human health and for natural reality that is partially expressed in contemporary environmental ethics. This study argues that the autonomy perspective in Christian ethics can integrate the insights of Christian creation theology with a moral and ethical evaluation of philosophical, scientific and economic perspectives. It unapologetically argues for the priority of the freedom and dignity of the human person. However, it can conceptualise this in terms of continuity with physical reality as well as discontinuity. In that way it contributes in the public realm to the critiques and development of policy in the area of transgenic food. Theologically autonomy defends the significance of creatureliness and the role of the steward in the recognition of the status of the non-human creation. It is precisely because Christian theology is 'anthropocentric' and does not collapse the distinction between the human person and the non-human creation that it can argue for human responsibility for the non-human creation. This study brings the insights and methodologies developed in theological ethics to bear on the question of transgenics and public policy. It also recognises the theological concern to critique environmental theologies and ethics in the light of the biblical and systematic aspects of Christian theology.

CHAPTER 1

Transgenics in Science and Economics

Transgenic techniques have been seen as just another tool in the kit of a plant breeder; any risks or hazards associated with them are, in theory at least, calculable and confinable. It is presumed that this technology is both subject to technical evaluation and open to technical amelioration. Given that, it would appear that the evaluation of transgenic organisms in agricultural production should rightly remain the prerogative of the scientific community. 'Science', it is suggested, will ultimately and properly be the final arbitrator in the ethical controversy over transgenic food. Already the potential for hazards to accrue from these novel interventions has prompted scientists to examine not just the intended outcomes but the unintended and negative consequences to human health and to the environment. The scope of biosafety research has ranged from the very specific and empirical (the characterisation of pollen distribution patterns from transgenic corn for example) to the aspirational (whether this technology can move agriculture further down the road to sustainability).

In this chapter I will examine how scientific evaluations in fact raise ethical and social questions that are not descriptive but normative in nature and which are outside of the remit of the sciences. Transgenic crop plants which were bred to be sterile for example, raised significant ethical issues about the wilful destruction of subsequent cropping capacity in agricultural crops. This was particularly so where food security is already subject to several threats. The issue is not whether the technology is effective but whether it is appropriate and for whom.

Technical problems with the first wave of transgenic organisms are considered to be either under control or at least under surveillance. Rather than a moratorium, increased regulation is the price being paid, by all farmers, for the increased industrialisation of the farming environment.

Regulation, it is assumed, will allow a level playing pitch for all growers, consumers will be allowed to exercise their preferences, and the market will decide. Yet from an economic perspective, the risks GM crops pose to the markets of non-GM crops, particularly in Ireland and perhaps in Europe, may simply not be offset by the modest agronomic advantages these crops achieve.

'Cost–benefit analysis' and 'risk–benefit analysis' are economic tools adopted by environmental managers to ameliorate the negative impact of agricultural technologies. The problem with the institutionalisation of cost and risk–benefit analysis is that it has the effect of already framing its conclusions. Economic valuation then becomes 'normative' for every value, a given to which we must conform. Yet, environmental goods are 'public goods' and in considering them institutions need to examine not just market choices, preferences and 'externalities', but also ethical and social principles. Consequently the utility and limitations of economic strategies in environmental management need ethical evaluation.

The concept of sustainability as a mediating principle between scientific evaluations and ethical and social perspectives is then outlined. It has both a descriptive aspect that is firmly based in the natural sciences and a normative aspect that needs ethical and social foundation. It has therefore ethical potential to integrate divergent viewpoints, and invites ongoing ethical elaboration.

The contention over the nature and extent of the risks and hazards from released transgenics persist, and may intensify with further innovation. It is a task for scientists and researchers to address the lacunae in the scientific record, to describe the outcomes for human health, and the environmental consequences of new introductions. Likewise it is a task for economics to indicate how economic choices in the use of this technology could contribute to scientific and social goals such as food security, sustainability and the 'coexistence' of GM and non-GM cropping in the EU and on world markets. In this chapter I will first broadly describe the discoveries and techniques, the intended and presumably desired outcomes for agriculture and the wider environment, as well as the undesired technical problems, associated with transgenic organisms. I will then examine the strategies that environmental managers have adopted; namely the regulation of novel foods and agricultural produce and the modification of cost–benefit analysis in relation to environmental protection. Finally I will examine the claim that transgenic technology, in its current state of development, is capable of delivering a more sustainable agriculture.

Developments in transgenics

The industrial use of transgenic micro-organisms is already a reality, as is the commercial production of pest- and herbicide-resistant crops in agriculture. Although genetically modified crops occupy a relatively small proportion of the world's agricultural acreage the area has been increasing steadily.

> During the nine-year period from 1996 to 2004, the global area of GM crops increased from 1.7 million hectares to 81.0 million hectares. This equates to 5% of the 1.5 billion ha of global cultivatable cropland. In 2004, GM crops were cultivated by some 8.25 million farmers in 17 countries, with a 20% increase in the area cultivated to GM crops from 2003 to 2004. Almost all (98%) of this was grown in only six countries: USA (59%), Argentina (20%), Canada (6%), Brazil (6%), China (5%) and Paraguay (2%). Other countries with significant plantings include India (1%) and South Africa (1%), while Uruguay, Australia, Romania, Mexico, Spain and the Philippines have less than 1% each.[1]

Additional transgenic innovations are following fast on the heels of these introductions: the nutrient enrichment of staple crops, the development of 'functional foods', vaccine farming in terrestrial crops, and far-reaching developments with terrestrial and marine organisms. Marine transgenics, it is suggested, will salvage commercial fishing through the replenishment

---

1 Department of Agriculture and Food, 'Co-Existence of GM and Non-GM Crops in Ireland', Department of Agriculture and Food, http://www.agriculture.gov.ie/index.jsp?file=publicat/publications2005/gm_coexistence/index.xml.

of fish stocks in the wild and through the exploitation of new feeding grounds by novel 'species'.

With these introductions comes the concomitant need for comparative data, to allow the evaluation of risks and hazards to environments from the release of transgenic organisms. Yet there are vast lacunae in current understandings of natural ecosystems. This need has led to investment in base-line studies of terrestrial and marine ecosystems, an unforeseen spin-off from biotechnology for a hitherto largely ignored sector. In addition, irrespective of their applied or industrial uses, DNA transfer techniques are significant new scientific tools, allowing groundbreaking studies of basic plant and animal biology.[2] These studies may in turn have further industrial applications in the field of biomimetics; the imitation of nature by nano 'machines' or nanotechnology.[3]

From the outset, even given the lacunae in the scientific record, researchers have acknowledged that there are foreseeable consequences from this new technology that potentially put human food security and environmental safety (collectively biosafety) in jeopardy. This has already had consequences. The reliable segregation and labelling of transgenics is demanded at least in the EU, ostensibly in the interests of consumer choice and to a lesser extent to safeguard the interests of non-GM growers. The demand for segregation has prompted the further development of 'genetic' containment techniques which in turn bring further technical difficulties and ethical questions. There has been an assumption that technical solutions can be found for the technical problems these techniques will engender.

There has been less scrutiny, from the perspective of science and economics, of the achieved agronomic advantages of novel introductions to date. The claims that transgenics have the capacity to move agricultural production to a more sustainable phase are asserted rather than tested.

---

2   P. Lurquin, *The Green Phoenix; a History of Genetically Modified Plants* (New York: Columbia Press, 2001), 112.
3   Nanotechnology is the branch of technology that deals in dimensions and tolerances of less than 100 nanometres. It is concerned with the manipulation of individual atoms and molecules.

Biotechnology companies have argued that genetically engineered pest and disease resistance in crops could prevent some of the environmental damage caused by conventional intensive farming techniques. Yet pest resistance build-up in monocultures may prove to be as much a problem for transgenic crops as it did for conventionally bred crops. Whether ongoing resistance to pests or improved nutrition or yield are achievable realities with current transgenic practices is still a debated technical point and not a foregone empirically tested conclusion.

I will consider, in the following sections, the science of recombinant DNA techniques, the technical 'breakthroughs' and some of the foreseen consequences, both intended and unintended, of the introduction of transgenetics into the agricultural environment. Relevant to the discussion is a description of some of the transfer techniques that have facilitated the production of transgenic organisms and the application of these techniques in the development of new micro-organisms, plants, arthropods, fish and mammals, for novel – and it is presumed useful – agronomic traits. Transgenic micro-organisms, plants and animals are currently under development if not already in commercial production. I outline the development and possible benefits of this technology, alongside the evidence for some of the technical risks and hazards to human health and the wider environment that could result from its adoption.

Transgenic techniques and applications

It has been argued that highly bred agricultural crops are, and have long been, 'genetically modified'; chosen for their superior genotype (genetic constitution) which is reflected in some ways in their phenotype (observable characteristics). Most of the world's staple crops have undergone dramatic cycles of modification through crossing and selection by farmers and breeders over many centuries. With the advent of DNA transfer technology 'genetically modified' has come to refer to organisms whose genetic structure is altered through DNA transfer and recombination between non-interbreeding organisms. Genes can be identified, isolated and reconstructed across species, genera and kingdoms, to give some agronomic advantage that could not occur using traditional selection and breeding programmes.[4]

There are several techniques involved; the most successful one, for plants, involves using a bacterium as a vector in transferring the novel genetic traits. 'Agrobacterium tumifaciens-mediated gene transfer' is the most widely used technique in transgenic plant production.[5] This technique was used in the production of insect resistance cotton (BT cotton). Lurquin, Professor of Genetics at Washington State University, narrates the history of the development of this versatile technique and other techniques such as 'biolistics', or gene guns.[6] This technique has several advantages over others, in terms of stability and predictability, compared with micro-mechanical direct gene transfer. It is less likely to result in complicated foreign DNA integration patterns or to other complications such as 'gene silencing'.[7] However, it does not allow the transformation of genes carried in cell organelles other than the nucleus; in mitochondria or chloroplasts for example, because the transfer-DNA (t-DNA) is driven to the nucleus. Where the transformation of chloroplasts and mitochondria is sought, for the production of medical and industrial products by plants, other techniques would have to be employed. In the case of animals for example the 'nuclear transfer technique' has been employed in the generation of transgenic sheep, goats, cattle, swine and poultry.[8]

The terms employed to describe the transformed organism – genetically modified (GM), transgenic, transformed, genetically engineered – do not however just refer to transfer methods. They do refer to the

4 G.L. Fletcher, S.V. Goddard and C.L. Hew, 'Current Status of Transgenic Atlantic Salmon for Aquaculture' (paper presented at the Proceedings of the 6th International Symposium on the Biosafety of Genetically Modified Organisms, University of Saskatchewan, 2000), 179.
5 Lurquin, *The Green Phoenix*, 2.
6 Cf. Ibid., 56–102 & 96–99.
7 Ibid.
8 P.S. Silva and B. Buchanan, 'Regulatory Considerations for Transgenic Animal Health and Food Safety Assessment; a Canadian Perspective' (paper presented at the Proceedings of the 6th International Symposium on the Biosafety of Genetically Modified Organisms, University of Saskatchewan, 2000), 187.

recombinant techniques employed but also to the source of transferred material. Strictly genetic material could be sourced from within the same species but more often it has been taken from other species, genera or kingdoms.⁹ These terms are used here to refer both to organisms altered by such techniques and to novel organisms transformed with genetic material sourced from outside the crop species, or their sexually compatible relatives. The debate over this technology has blurred the distinctions (between kingdom, genus and species for example) traditionally so confidently made and codified by taxonomists, dating back to for example Albertus Magnus in the medieval period and to Linneaus in the eighteenth century. Many researchers argue that there is a qualitative difference between the products of these techniques and those of traditional breeding programmes. Others suggest that the taxonomic system does not represent fixed and unchanging species and genera but is merely a human construction which is a heuristic devise that allows prediction.¹⁰ Yet, the taxonomic system may represent more than just a human construct or a projection. It maps variability in nature and it refers to something empirically significant and verifiable; barriers to crossing exist, and may do so because they confer some fitness advantage. Transgenics in this view may not be a technique that simply speeds up a slow process, it could constitute a whole different way of breeding crops. There is, however, no clear consensus among the scientific community over the significance of taxonomic distinctions in relation to transgenics, let alone what that difference consists in or what the significance of considering it

---

9   The emphasis on transcending species barriers may prove to be the more trivial contribution to altering crop plants in a predictable and advantageous way. The techniques may prove more useful in enhancing breeders' capacity to explore and exploit gene pools within crop species. Cf. E.J. Walsh, 'Genes, Plants and Mankind: A Plant Breeder's Perspective on the Reality and Potential of Plant Biotechnology' in *Research Report of the Faculty of Agri-food and the Environment* (Dublin: University College Dublin, 2004), 82.
10 Cf. Alex Rosebergs, *A Philosophy of Science* (London: Routledge, 2000), 70.

might be. This tension has been articulated in many ways, some of which will be referred to below.[11]

Some foreseen but unintended consequences

There are unintended side-effects associated with the use of transgenic organisms and much of the controversy over recombinant DNA technology has centred on the negative consequences that might flow from its application. Transgenic organisms, it is suggested, may pose immediate risks to human health and long-term risks to the environment.

The risks to human health through the consumption of transgenic food have been assessed, almost exclusively, using the concept of 'substantial equivalence'. The relative safety of novel foods is adjudicated on the basis of their 'similarity' to known foods.[12] However, the concept of substantial equivalence is considered more a regulatory tool than a bona fide scientific principle that can predict or help mitigate risks.[13] To date all the evidence suggests that there are no negative side-effects, in terms of immediate human health, from the consumption of transgenic foods. It is more correct to say, however, that there is no known risk to human health from the agricultural crops that have been cultivated to date worldwide. These crops appear to have a history of safe use, as regards human health, having undergone thorough food safety evaluations.[14] Nonetheless since there is no system for monitoring, tracing, or documenting exposure

---

11 Paradoxically the biotechnology industry argues both that transgenics are no different to conventionally bred crops and therefore pose no risk to consumers, but are so different as to warrant a change in the rules governing the patenting of living organisms. Cf. E Millstone, E. Brunner, and S Mayer, 'Beyond "Substantial Equivalence"' *Nature* 401, October (1999).
12 This was first introduced in 1993 by the Organisation for Economic Cooperation and Development (OECD) and is now referred to in common parlance as natural equivalent.
13 Millstone, Brunner, and Mayer, 'Beyond "Substantial Equivalence"'.
14 Walsh, 'Genes, Plants and Mankind: A Plant Breeder's Perspective on the Reality and Potential of Plant Biotechnology', 80.

to transgenic food with widespread use this is a description unlikely to develop under present assumptions.

There are nonetheless some issues that need to be monitored. Lurquin suggests that transgene expression may lead to the production and build-up of harmful secondary metabolites in a host that would otherwise appear phenotypically as normal.

> Many are concerned that random integration of transgenes may have cryptic effects on the host plant – that is, undesirable effects with no immediate phenotypic impact but with potentially harmful consequences for the consumers of this plant material. This could happen, for example, if the synthetic pathways of secondary metabolites were inadvertently disrupted by transgene insertion. Toxic compounds might then accumulate in the host.[15]

In the description of the problem of antibiotic resistance in transgenic plants, it is also noted that it is difficult to rule out the appearance of foreign gene products in some foods; foreign pollen in honey for example.[16] Rather than simply ignoring the possibility, ways can be suggested to ameliorate the potential for cryptic effects through improved gene targeting or removing suspect source genes altogether (for anti-biotic resistance for example). There is a growing body of research and literature on the production of 'nontransgenic products' from transgenic plants using genetic containment and elimination strategies.[17] This is designed to secure foods for human consumption. It may also be motivated by the additional desire to bypass labelling requirements for GM organisms (GMOs) in the EU.

Labelling notwithstanding, segregation also presents problems. There have been some high-profile documented examples of transgenic produce which was not passed for human consumption being found as

---

15 Lurquin, *The Green Phoenix*, 115.
16 C.N. Stewart, ed., *Transgenic Plants: Current Innovation and Future Trends* (Wymondham: Horizon Scientific Press, 2003), 121.
17 R.J. Keenan and W.P. Stemmer, 'Nontrangenic Crops from Transgenic Plants', *Nature Biotechnology* 20 (2002).

a contaminant in food that was passed for human consumption.[18] This raises issues concerning food segregation (beyond questions of segregation in the interests of consumer choice), the security of food supply lines and even marketing, since 'sustainable' and 'organic' farmers rely on traceability in drawing a premium price for their produce. Transgenic crops are considered unsuitable for 'organic' growing, for both technical and philosophical reasons, and grower organisations have objected to the threat to their livelihood from the adventitious (accidental and occasional) presence of transgenes in organic and indeed in high-value non-GM produce.[19]

Beyond immediate human health issues, long-term risks and risks to the broader environment from transgenics are harder to foresee and, as a result, are more difficult to manage. This is especially the case in light of the lack of base-line studies with which to compare the consequences of new introductions. There has been a lot of research activity in some areas, most particularly in advance of and during commercial introductions. The risks and hazards arising from gene flow from herbicide-resistant transgenic crops to weedy relatives have been the subject of a considerable and growing body of scientific literature. For many operators these problems could be eliminated or ameliorated with further technical solutions like genetic containment strategies. Yet despite the evidence that the risk of gene flow is considerable (the likelihood is that transgenes will find their way into weedy relatives), the hazard (the likelihood that this will have negative consequences) is assessed as being low. This is considered the case even in areas where important landraces (traditional local varieties which remain an important source of variation for breeding programmes world-wide) are grown. The risk is assumed to be balanced by the benefits even though the benefits are again more asserted than established. So for many GM scientists, although we must avoid risk where possible

---

18 A Pollack, 'Plan for Use of Bioengineered Corn in Food Is Disputed', *The New York Times*, 29 November 2000.
19 A Bouchie, 'Organic Farmers Sue GMO Producers' *Nature Biotechnology* 20 (2002), 210.

or practicable, only well substantiated hazards should slow up the commercialisation of a transgenic organism.

It is clear that 'risk' is more and more invoked as a way of urging caution rather than being thought of as a significant probability of harm.[20] Risk is being reclassified and ethically neutralised in the literature. It does not translate as 'undesirable outcome' (or the possibility of such). This judgement is reserved for the category 'hazard' and it is hazards that must be avoided. This distinction is not always clearly made by researchers and can be the cause of some confusion. Not all risks are hazards but there are hazards that may be discounted because they are inadvertently classified as risks.

The foreseen risks and unintended hazards from some novel introductions of transgenics micro-organisms, plants and animals are examined below with particular emphasis on herbicide- and insect- resistant crops, and transgenic fish.[21] Transgenic terrestrial farm animals and transgenic mammals are not the focus of this study, and would warrant a separate study of their own. It is clear from this analysis that transgenic applications raise technical issues and problems which cannot be simply offset by further technical measures.

The transgenic modification of specific food organisms

In this section I outline some of the benefits claimed for this technology in the production of food organisms. This is followed by a short discussion of potential attendant negative outcomes, their likelihood and their impact. It seems clear from studies in the environmental biosafety of transgenic organisms that each novel introduction demands specific evaluation.

---

20 Paul Thompson, *Food Biotechnology in Ethical Perspective* (London: Blackie Academic and Professional, 1997), 229.
21 It is also necessary to go beyond listing the negative impacts of new technologies since stressing outcomes, in consequentialist mode, invites only the amelioration of practices rather than any system-wide reform. Paul Thompson, *The Spirit of the Soil: Agriculture and Environmental Ethics* (London: Routledge, 1995), 21.

The transfer of this technology to the biosphere requires elaboration of base-line studies, further evaluation of the presumed advantages of containment strategies, and long-term post-introduction monitoring and feedback measures. I specifically treat transgenic alterations to micro-organisms, plant crops and fish.

Micro-organisms

The first transgenetic organisms were micro-organisms. Their applications were generally confined to laboratory or industrial settings, as for example in the production of human insulin from the modified gut bacterium *Escherichia coli*. However there have also been releases into the wider environment, as with the so-called 'ice-minus' bacterium designed to protect fruit and flowers from spring frosts, with no adverse effects.[22] These led the way for further releases and was followed by over 2000 plant and micro-organism field applications in the terrestrial environment.

Certain micro-organisms have the known potential to counteract some of the negative impacts of adverse environmental events. They can act as bioremediators and bio-indicators, ameliorating negative pollution events or indicating alterations in ecosystems. They are capable of biodegradation, breaking down undesirable complex molecules, effectively removing them from the environment. However, naturally occurring micro-organisms have had limited application. There is some evidence that transgenic micro-organisms could facilitate more effective bioremediation – the deliberate application of 'biological activity' – to reduce the concentration of harmful contaminants.[23] Engineers hope to develop more effective organisms for bioremediation, to optimise breakdown enzymes and increase the range of activity and improve the efficacy of known natural bioremediation organisms.

22 M.A. Levin and H. Strauss, 'Overview of Risk Assessment', in *Risk Assessment in Genetic Engineering*, ed. M.A. Levin and H. Strauss (New York: McGraw Hill, 1993), 4.
23 R. Zilinskas and P. Balint, *Genetically Engineered Marine Organisms* (Boston: Kluwer Academic, 1998), 111.

Root bacteria play a major role in plant development and it is assumed that these micro-organisms could be transformed for the improvement of crop growth and development. 'Plant growth promoting rhizobacteria are of increasing importance for bio-fertilisation, bio-stimulation and biological control of plant pathogens in sustainable agriculture.'[24] Transgenetics could make further contributions in this area. In China transgenic microbial agents with reported nitrogen fixation properties (converting atmospheric nitrogen to a form available to plants as a nutrient, a process associated commonly with microbial symbionts of leguminous plants) have been field tested. These may have the potential to increase yields while reducing the need for nitrogen fertiliser.[25]

Transgenic micro-organisms may however pose risks to human health and the environment. The first considerations of risk to human health were raised in the United States of America in the 1970s when plans to clone an animal tumour virus in *E. coli* were abandoned.[26] It was feared that the new artificial recombinant DNA could transfer cancer gene characteristics to humans, since the altered organism could combine with the *E. coli* normally found in the human gut.[27] The use of antibiotic-resistant plasmids (bacterial DNA, separate from the chromosome) as vectors and markers in gene-transfer experiments was a second cause for concern.[28] It was anticipated that bacteria might acquire antibiotic resistance via

---

24 K. Smalla, 'Field Releases of Genetically Modified Micro-Organisms', *Environmental Biosafety Research* 2 (2003), 65.

25 S. Jia and Y. Peng, 'Research and Regulation on Biosafety of GMOs in China', *Environmental Biosafety Research* 2 (2003), 59.

26 Cf. J. Tomiuk and A. Sentker, 'History of and Progress in Risk Assessment', in *Transgenic Organisms: Biological and Social Implications*, ed. J. Tomiuk, K. Wohrmann, and A. Sentker (Basel: Birkhäuser Verlag, 1996), 217.

27 For this reason, Klaus Amman (Director, Bern University Botanical Garden) claims that the comparison made between the release of transgenic organisms and the introduction of pesticides is ahistorical. It was assumed from the outset that pesticides were unconditionally safe. Public Information Symposium, University College Dublin, October 1999.

28 Cf. P. Wheale and R. McNally, *Genetic Engineering: Catastrophe of Utopia* (Hemel Hempstead: Harvester Wheatsheaf, 1988), 43.

such plasmids residing in them. Although these risks were later downplayed, the issues have not disappeared. Since there are alternative markers available now and there has been continued pressure to reduce the use of antibiotics, there are moves to eliminate antibiotic-resistance marker use altogether, at least in the EU.[29] It is suggested, negatively, that this is a case of public pressure altering the 'trajectory' of research projects, even in the absence of evident 'risk'. Indeed, for Day the risk to human health from antibiotic markers used in transgenesis is far less significant than other antibiotic resistance factors we are already exposed to in the environment.[30] However, the move to eliminate these markers could be evaluated more positively. The dangers were identified such that other, apparently equally effective, methods are to be adopted where available, with no great loss to the research. There is even greater conflict in scientific research when the dangers that are to be endured or tolerated are considered 'unavoidable'. In that case all frontiers are to be crossed first and the casualties counted later. From an ethical perspective, so-called 'unavoidable' techniques cannot be used to justify harms from deliberate human actions.[31] The potential to substitute with less risky techniques should be a welcome development.

Plants

Crops engineered for herbicide or insect resistance are already widely grown in the Americas, China and India. These particular innovations have the potential to contribute to pest control management across a broad range of crops and agronomic situations.[32] Built-in bioengineered resist-

---

29   Anil Day, 'Antibiotic Resistance Genes in Transgenic Plants: Their Origins, Undesirability and Technologies for Their Elimination from Genetically Modified Crops', in *Transgenic Plants: Current Innovations and Future Trends*, ed. C.N. Stewart (Wymondham: Horizon Scientific Press, 2003).
30   Ibid., 126.
31   Thompson, *Food Biotechnology*, 231.
32   'The development of transgenic crop plants expressing toxins from the soil bacterium *Bacillus thuringiensis* for managing insect populations represents one of

ance in crops brings several agronomic advantages. It reduces the need for insecticide use in intensive farming systems. This could contribute to the survival of non-target organisms, increasing in-crop biodiversity. Since many non-target organisms are also pest predators, there is a further improvement in pest control over conventional farming methods with their survival.[33] This could have the added benefit of reducing the levels of insecticide residues in food and water and decreasing pesticide risks to farm workers in pesticide-intensive cropping systems.

Plants could be transformed with other objectives in mind. There is potential for engineering agronomic traits in plants such as metal tolerance, drought resistance and improved nutritional status; as in the much publicised case of vitamin-A-enriched rice which could contribute to the health of marginalised populations. Transgenic technology could have applications in the production of industrial polymers, pharmaceuticals and even vaccines.[34] The implication is that edible vaccines, produced in plants engineered with human pathogen genes coding for proteins that trigger an immune response in infected individuals, would not require syringes and hence no sterilising equipment or refrigeration. Plants could also be modified to survive in the presence of, sequester and even degrade organic

---

    the most significant changes in pest management practices since the advent of chemical pesticides and is likely to have profound effects on future agricultural production. The range of crops that have been shown to be capable of genetic transformation now includes many of the major food and fibre crops, and the history of this technology over the past decade suggests that essentially all major crops should be amenable to transformation'. B.D. Siegfried, 'BT Transgenic Plants for Pest Management; Challenges and Opportunities' (paper presented at the Proceedings of the 6th International Symposium on the Biosafety of Genetically Modified Organisms, University of Saskatchewan, 2000), 113.

33 'The adoption of BT cotton allowed insecticide spraying to be reduced by 70% and 80% (from 15 to 2 times on average). Due to the reduced use of chemical pesticides, the biodiversity of insects and arthropods has been increased in the BT cotton field. The increased natural enemies (predators in particular) also efficiently prevented cotton aphid from resurgence that would have occurred in non-BT cotton field with insecticide application.' Jia and Peng, 'Research and Regulation on Biosafety of GMOs in China', 58.

34 Stewart, ed., *Transgenic Plants: Current Innovation and Future Trends*, 2.

and heavy-metal pollutants.[35] This could stabilise damaged environments and even, it is suggested, open up new, previously inhospitable areas for crop production.

Transgenesis has also become an important tool in the understanding of basic plant biology pathways, in plant development, differentiation and gene regulation.[36] This has further implications in other spin-off disciplines; bioremediation and the development of biomonitoring devices for example.[37] Although such developments are currently more science-fiction than science-fact, they are envisaged as 'nano-machines' like 'super-bacteria' capable of disassembling organic wastes and reconstituting new valuable molecules.[38] A whole new area of biomimetics, or biomolecular technology awaits: the production of semi-synthetic 'machines' that imitate or mimic 'biologically programmed molecular recognition' systems like that of the ribosome itself.[39] The ribosome is envisaged as a natural cellular 'nanomachine', an information template capable of the manipulation of atoms and molecules with dimensions and tolerances less that 100 nanometres (10^{-9} or one thousand millionth of a metre).[40] This basic research, it is suggested, satisfies a need to better understand the molecular basis of life and motivates research to 'emulate natural systems in order to produce artificial devices of entirely novel functionality and performance'.[41] This could have applications in, for

---

35   Lurquin, *The Green Phoenix*, 118.
36   Ibid., 112.
37   Cf. W. Verstraete, 'Environmental Biotechnology for Sustainability', *Journal of Biotechnology* 94 (2002).
38   Ibid., 96.
39   A ribosome is a non-membranous cell organelle composed of protein and RNA that is the site of protein synthesis, and is found in the nucleus, cytoplasm, mitochondria and plastids of a cell.
40   C.M. Niemeyer, 'Semi-Synthetic Nucleic Acid–Protein Conjugate; Applications in Life Sciences and Nanotechnology', *Reviews in Molecular Biotechnology* 82 (2001), 48.
41   Ibid.

example, drug delivery systems, light-harvesting devices and in devices that interface electronic and living systems.[42]

Several risks however have been associated with transgenesis in plants. The use of transfer DNA to confer insect resistance brings with it some of the same problems identified by integrated pest management systems in 'chemically' dependent agriculture. Pest populations can resurge and producers are forced back onto the insecticide 'treadmill' to control the problem.[43] The transfer of transgenes (gene-flow) and the stable establishment of genes (introgression), in weedy relatives of important food crops, has been considered a risk with transgenic crops. Crop plants share a history with their wild relatives. The risk of gene-flow and introgression is more likely than could at first be established. Studies 'illustrate that containment of transgenes is probably unrealistic for most commercially grown crops, and even a moderate degree of confinement may be difficult to achieve'.[44] The focus now is on whether this substantial risk presents a hazard to human health and the environment. Gene flow would present a hazard, it was suggested, where the spread of herbicide resistance to weedy relatives would result in the emergence of new weeds that are even more difficult to control, necessitating the increased use of herbicide with a concomitant 'biodiversity' impact. This could be exacerbated in the case of 'gene-stacking' where the occurrence of several new traits in 'wild' progeny (weeds) may result in an accelerated adaptation of insects, weeds and diseases as well as unintended effects on non-target organisms.

In plants, gene flow has mostly been considered pollen mediated and restricted to short distances, although hybridisation has been reported hundreds of metres away from cultivated sources. Generalisations in hybridisation predictions are not straightforward however, and dispersal is also possible via seeds and vegetative propagules. Better characterisations

---

42   Ibid., 62.
43   Thompson suggests that the debate over transgene-induced tolerance reprises the debate over pesticide use. Cf. Thompson, *The Spirit of the Soil*, 40.
44   A. Snow, 'Consequences of Gene-Flow', *Environmental Biosafety Research*, no. 2 (2003), 44.

of seed dispersal and persistence is needed.⁴⁵ However, it seems clear that the rate of gene flow and introgression does depend on the 'architecture of reproductive barriers between species' and the relative 'cost' in fitness the crop trait imposes.⁴⁶ Some traits may reduce the fitness of free-living relatives – the production of pharmaceuticals most likely would – while others may confer a fitness advantage. The consequences of crop-to-weed gene flow should therefore be considered as crop- or variety-specific, site-specific or even season-specific.⁴⁷

There is more and more evidence to suggest that there is a substantial risk of gene flow; the question remains as to whether this risk can be evaluated as a hazard. As with insect resistance, agronomic advantages gained with a reduction in herbicide use would be quickly eroded by the survival of herbicide-tolerant weeds. This has occurred and has had a negative effect on farming. In the US, the use of Roundup (a long-standing proprietary herbicide), has led to the build up of herbicide resistant weeds on agricultural land. As Walsh reports; 'US realtors specialising in the sale and renting of farmland have recently reported that the presence of Roundup-resistant weeds on a farm property reduces its rental value by about 17 per cent'.⁴⁸ In this instance an environmental risk has been neutralised by being translated into an economic hazard and the costs transferred to the farmer or land user. The assumption is that economic trade-offs can always be made. What is not always explicit is who bears the cost of these 'externalities'.

The possible 'contamination' of landraces has also been a subject of much concern in the biosafety literature. Landraces are globally significant sources of variation for plant breeders.⁴⁹ Plant breeders place great

45   E. Jenczewski, J. Ronfort, and A. Chevre, 'Crop-to-Wild Gene Flow, Introgression and Possible Effects', *Environmental Biosafety Research* 2 (2003), 12.
46   Ibid., 15–18.
47   Ibid., 19.
48   Cf. Walsh, 'Genes, Plants and Mankind: A Plant Breeder's Perspective on the Reality and Potential of Plant Biotechnology', 81.
49   Broadening the genetic base of crops has become an objective in light of the progressively narrower genetic base that agriculture depends upon.

significance on broadening the genetic base of agricultural crops, to protect in particular from the overdependence on monocultures, which can be vulnerable to large-scale devastation.[50] The protection of landraces, as a resource for world agriculture, is usually considered in the context of biodiversity. Maize landraces in Mexico, for example, contain transgenes from imports of maize from the US. These imports are not segregated. The indigenous practice of Mexican farmers, of using introduced varieties as a source of biodiversity, is extended to GM crops, whether deliberately or inadvertently.[51] For Mexican farmers gene-flow appears to be welcomed as an addition to biodiversity. Yet, it is unclear how GM introgression into landraces can be evaluated as regards biodiversity. It could be claimed that GM introgression 'adds' to biodiversity. However, as Walsh comments this is so only in a trivial sense.

In a counter-move, aimed at protecting from introgression, and it must be said to protect market share, some biocontainment strategies have been considered by biotechnology companies, to reduce the potential spread of transgenes.[52] However, these strategies in turn have raised further technical and ethical issues. The biological containment strategy known as 'terminator technology' was developed to ensure crops from transgenic seed did not produce viable seed in turn. This built-in obsolescence was specifically aimed at protecting seed market share, particularly in poorer developing countries where the infringement of proprietary rights was deemed likely. However, relying on this kind of seed would require farmers, in the long-run, to either keep separate seed crops or to return to the seed company for future crop seed, since they would not be in a position to save it themselves. This was not well received by farmers and producer groups, particularly in the developing world; the very

---

50 H. Cooper, C. Spillane, and T. Hodgkin, eds, *Broadening the Genetic Base of Crop Production* (CABI, 2001).
51 A. Avarez-Morales, 'Possible Implications of the Release of Transgenic Crops in Centres of Origin or Diversity', *Environmental Biosafety Research* 2, no. 47–50 (2003).
52 J. Schiemann, 'New Science for Enhanced Biosafety', *Environmental Biosafety Research* 2 (2003).

markets seed companies found themselves vulnerable in, and towards which biocontainment was aimed. It raised significant ethical issues about the wilful destruction of cropping capacity – ostensibly to protect seed market share – in the context of already precarious subsistence farming. Containment strategies protect the interests of seed companies more than they do food security or farm environments. They are at odds with the claim that this technology will feed the world.

The issue of containment is crucial in the transformation of plants for the production of vaccines or in the 'biopharming' or 'molecular farming' of industrial chemicals. There are potential risks for the environment and the consumer; such products may be more akin to medicine than to food and would need to be treated accordingly.[53] In terms of human health, the development of these so-called second generation transgenics raises questions about the wisdom of mass-medication through the food chain and the efficacy of 'functional foods'. These foods are designed to contribute health benefits over and above their normal nutritional capacity, whether GM derived or not. The distinctions between food and medicine are being blurred. In the case of transgenics, gene flow, and the introgression of 'nutraceutical genes' to food-crop relatives, could pose significant and foreseeable hazards. Researchers are faced with considerable tasks. Isolating the transgene and its products may best be achieved by introducing it into a non-food crop and extracting it later. Applying less absolute biocontainment strategies may risk introgression into food crops, with hazardous consequences. It seems clear that transgenics of this type do not lend themselves to being farmed in the conventional sense.

Lastly, transgene-induced herbicide- and insect-resistance in crops could have positive benefits as regards pesticide use. They could reduce the impact of pesticides on wildlife in intensive-farming environments. The claim that transgenic modification would help in a shift away from the use of more contaminating pesticides could be true; if for example, the pesticides used in their production replaced persistent residual pesticides. Yet there is some evidence, from conventional systems, that

---

53   Ibid., 41.

even with replacement, this outcome is not as likely as might be at first presumed. Indeed although the volumes and rates of active substance for pesticides have been falling since the early 1980s in the UK, the rate of decline in farmland wildlife over that period has continued to increase. Although more than one factor has contributed to this decline, there is some evidence that the efficacy of agrichemical use in terms of targeting, specificity and timing plays a role.[54] It is not just the input in terms of volumes that needs to be tracked. The overall impact of specific regimes also needs monitoring. Only a wider assessment of impacts will allow regulators to examine the full implications of the commercial introduction of transgenic cropping systems on biodiversity. Current transgenic crop introductions only offer symptomatic relief from 'chemically-dependent' farming rather than a substantial alternative to input-intensive systems.

> Combinations of these traits may eventually offer us options of low- or no-chemical input crop systems, able to co-exist with native biodiversity far more successfully than the current unsustainable systems that are continuing to damage farmland biodiversity throughout the world. This goal of greater sustainability will only be reached if protection and enhancement of native biodiversity and the wider environment are written into research and development programs as specific objectives, alongside those of better agronomic performance and better food quality.[55]

A broader measure of 'sustainability' in farming would be required to assess their relative utility. Sustainability, as a laudable goal for agriculture, cannot simply be co-opted for a production system that can claim only a modest reduction in herbicide use.

---

54 B. Johnson, 'The Environmental Impact of GMOs: Some Possible Problems and Some Possible Solutions' (paper presented at the Proceedings of the 6th International Symposium on the Biosafety of Genetically Modified Organisms, University of Saskatchewan, 2000), 40.
55 Ibid., 44.

Fish and marine organisms

There is an acknowledged threat to natural fish stocks from the overexploitation of commercial fishing. Current fish farming techniques, it is suggested, will not make up the shortfall and in fact may be exacerbating the problems. The introduction of fishing quotas in the EU, and other world-wide measures to curb over-fishing, anticipate the need to reduce the fish-take to rebuild stocks. Transgenics is heralded as one possible mechanism that could offer a technical solution to over-fishing.

> In light of the assumption that world fisheries are over-exploited and commercial fishing endangered, aquaculture needs to move into a new phase; in particular the development of genetically superior stocks with improved, for example growth rates, disease resistance and tolerance of biotic and abiotic survival and growth limiting factors.[56]

Improved overwintering of salmon by inducing freeze-resistance, would allow the exploitation of new waters for feeding, and the introduction of a growth hormone gene would accelerate growth rates. It seems that

> ... the majority of marine transgenic research has focused on creating organisms that are optimal for aquacultural harvesting, possessing characteristics such as large body size, increased growth rate, greater disease resistance, and greater ability to survive adverse environmental conditions relative to the parental species.[57]

Marine biotechnology has already added foreign genetic material to at least 14 species of fish and several species of molluscs and crustaceans, with many being engineered for higher yields and faster growth rates.[58]

Particular potential risks and hazards present in the area of marine transgenetics. Some spring from consideration of the unknowns in marine biology, some from the nature of the marine environment itself, and some

---

56 Fletcher, Goddard, and Hew, 'Current Status of Transgenic Atlantic Salmon for Aquaculture', 179.
57 Zilinskas and Balint, *Genetically Engineered Marine Organisms*, 67.
58 Levin and Strauss, 'Overview of Risk Assessment', 10.

from the characteristics of specific transgenic organisms. Risk assessments are rendered even more provisional in light of the scale of unknowns in marine biology, the lack of base-line studies and the particular attributes of marine organisms.[59] The oceans represent over 90 per cent of the total volume of the biosphere, the remainder being land and air, and contain 97 per cent of water in the hydrosphere.[60] These are attributes of the marine environment that complicate risk assessment and management issues. Scale is not irrelevant and although 'on a macro scale, the questions to be asked when evaluating ecological risk are similar for terrestrial, freshwater, and marine environments' the ecological risks with transgenic organisms are higher in the marine environment, in part because of the potential scale of the risks and hazards.

Muir and Howard, marine biologists, identify a qualitative difference between transgenic and conventionally bred organisms that has particular implications for marine transgenics. They qualify this difference in terms of pleiotropic effects; where a single gene produces two or more apparently unrelated effects. Transgenesis discounts these pleiotropic effects.

> While selective breeding is based on polygenic inheritance whereby the result is the cumulative effect of many (perhaps hundreds) of genes each with small effect; in contrast, transgenesis involves one gene with a major effect. An important consequence of this difference is the pleiotropic effects of the genes involved. In the selective breeding process, all the correlated traits that support enhanced growth and reproduction, such as skeletal and vascular systems, are also selected; this is not the case in the production of transgenics.[61]

Their discussion of the risks and hazards of transgenic fish is predicated on the possible pleiotropic effects of transgenes on marine organisms. The 'insertion of growth hormone gene constructs in fish is of concern because accelerated growth (particularly in combination with attainment

---

59 Zilinskas and Balint, *Genetically Engineered Marine Organisms*, 88.
60 Ibid., 32.
61 W. Muir and R. Howard, 'Assessment of Possible Ecological Risks and Hazards of Transgenic Fish with Implications for Other Sexually Reproducing Organisms', *Transgenic Research* 11 (2002), 108.

of large adult size) has many pleiotropic effects'.[62] The implication is that accelerated growth rate may adversely effect other fitness criteria, such as susceptibility to disease, or fertility.

Transgenic fish may also present particular problems for marine environments because of their particular characteristics, not least their high dispersal ability, low probability of recapture and high reproductive potential. So although aquatic plants may be relatively easy to control (although this is by no means established), this assumption is not justified in the case of introductions of micro-organisms or other marine organism. 'As in the terrestrial environment, it may be relatively easy to contain, monitor, and ultimately eradicate introduced aquatic plants. In contrast it may be impossible to contain or later eliminate micro-organisms once they are introduced into the open marine environment' and species of shellfish and finfish would fall somewhere between these two extremes.[63] A much greater understanding of how transgenics have ecological impact is needed if it is to be possible to develop risk and hazard assessment protocols that can be used with confidence in marine aquaculture.

Muir and Howard identify the need to develop an agreed risk assessment methodology.[64] They identify two particular harms (hazards) as opposed to risks with the introduction of transgenic fish; the local extinction of species, and ecosystem disruption. The first hazard, local species extinction, may occur where the transgene alters 'net fitness components'. Transgenic fish might for example confer a mating advantage that ensures transgene-flow into local populations, but at the same time reduce the number of offspring surviving to sexual maturity.[65]

The second hazard they identify could result from an effect that has been seen before with non-GM species introductions and resultant invasions. A pattern of biological invasion can be built from past introductions of novel species to new environments. A conservative pattern of biological invasions can be described as follows; only 10 per cent of

---

62   Ibid., 109.
63   Levin and Strauss, 'Overview of Risk Assessment', 10.
64   Muir and Howard, 'Assessment of Possible Ecological Risks and Hazards', 101.
65   Ibid., 102.

exotic species introductions succeed and of these only about 10 per cent produce adverse effects. The problem is that where they do establish themselves invasive exotic species can cause dramatic ecological disturbance.[66] Transgenic fish, Muir and Howard suggest, have the potential to create an extinction hazard or disrupt ecosystems. Prior to any introductions therefore, a system-wide integrative approach is needed to predict risk and hazards.

Aquatic biotechnology research has been limited principally to the laboratory or to quarantine ponds and little is known of the impact of introductions in the open marine environment.[67] 'Genetic containment strategies' have been proposed for marine transgenics to reduce the spread of a transgene in wild populations and to allow further research on biosafety issues. Reducing male fertility in the altered organisms is one such strategy. However, although containment is desirable, containment measures may also bring their own risks and hazards. For example there is some suggestion that reducing male fertility, without eliminating it, may increase the extinction hazard for wild species.[68] Introgression of transgenes may occur to non-GM organisms and the genetic containment effect – a reduction in fertility – be transferred unintentionally to wild species. In consequence some alternatives to widespread release have been proposed. Aquaculture development could take place on land, in high-quality recirculating water systems; this could give growers greater control over disease, parasitic infections, feeding regimes, temperature and photoperiod. This would allow the 'full' domestication of farmed fish while going some way to addressing the biocontainment question.[69]

It appears that given the scope of the unknowns in marine transgenics, researchers are more reluctant to promise too much for this technology, particularly in the short term. There are many problems associated with the genetic manipulation of terrestrial animals also; the reactivation of

66 Levin and Strauss, 'Overview of Risk Assessment', 11.
67 Ibid., 10.
68 Muir and Howard, 'Assessment of Possible Ecological Risks and Hazards', 107.
69 Fletcher, Goddard, and Hew, 'Current Status of Transgenic Atlantic Salmon for Aquaculture', 182.

latent diseases, for example, is a 'technical hitch' that puts the introduction of xenotransplantation (animal to human organ 'donation') on hold. Disease risks and hazards are not isolated to, but are multiplied by, the development of transgenic animals; not least the threat of the reactivation of retroviruses found in animal genomes. This is not helped by the fact that detection strategies for unanticipated agents are limited. Transgenic animals could risk the 'disruption of normal immune mechanisms by gene insertion, crossing of species barriers by pathogens and activation of latent viruses. Potential transfer of infectious agents during in vitro genetic manipulation procedures may give rise to the possibility that transgenic animals may act as sources of novel disease entities or disease agents with enhanced virulence attributes'.[70] Transgenics also raises many other ethical questions in relation to the welfare of farmed and other animals, apart from the consequences for human health.

Yet, for the most part, all of the negative outcomes mentioned above are considered technical in nature and so in theory open to amelioration or elimination by technical means, or using market mechanisms. Rather than a continued moratorium on releases pending further research (a strategy recently endorsed by Switzerland), in most European jurisdictions, in the European Union for example, the policy strategy in relation to transgenics has become one of regulation. Regulation aims to contain any risks and hazards that might accrue, either to human health or to the environment, with transgenic releases. It aims to serve so-called 'consumer choice', in relation to transgenics, through a product labelling system for already-approved GM crops and foods. The focus in European policy making has also and recently (since 2003) undergone a further shift. The focus in agriculture is no longer on the risks and benefits of transgenic crops but on the question of facilitating the 'co-existence' of GM and non-GM crops in member states. EU-approved transgenic crops are presumed to be 'safe' in terms of human health and the environment; although strictly, they present known risks, in terms of gene-flow, but

---

70   Silva and Buchanan, 'Regulatory Considerations for Transgenic Animal Health and Food Safety Assessment; a Canadian Perspective', 187.

no detected hazards, to either consumers or farm environments. Based on that presumption, regulation aims to allow for not only 'consumer choice' but also so-called 'farmer choice'. Each member state is to develop its own regulatory framework – under the principle of subsidiarity – to allow for the 'co-existence' of approved GM and non-GM (the latter to include conventional and organic) cropping systems.

From an ethical perspective however, regulatory frameworks represent a *de facto* decision in favour of using this technology. They cannot in themselves deliver on biosafety. The utility and the ethical limitations of regulatory and economic tools need further evaluation.

Strategies in environmental management

Several 'partial' policy strategies are used to limit or ameliorate some of the unintended, but foreseeable, consequences of recombinant DNA technology. Two such interdependent strategies are broadly outlined and assessed here; regulation, and cost–benefit analysis. Regulation will be broadly considered in relation to the following: food safety and the notion of substantial equivalence; the question of regulatory 'triggers'; the precautionary principle and the Cartagena protocol.

I will then examine the role of cost–benefit analysis and risk–benefit analysis in environmental management and suggest that these measures are not themselves value neutral. They may provide a way of organising information but as institutional measures they do not substitute for ethical principles or for an evaluation of the 'public good'. It is clear that some goods are not commensurable. A clean environment is a public good, and more than just a commodity. It is a context of life.[71]

---

71 In chapter two I take up the related concept of the common good, through which theological evaluations of the public good are articulated.

The regulation of transgenic organisms

Regulation, where it exists in relation to transgenics, attempts to manage all of the outcomes of novel transgenic introductions. Where policy makers anticipate the widespread introduction of transgenics, and all the economic and other benefits that should accrue, regulation appears to responsibly offset any uncritical or potentially harmful enthusiasm for this new technology. On the other hand, given the irreversibility, the potential negative outcomes of invasions and the hidden potential for the transfer of costs (to growers or consumers or the environment), regulation could be interpreted as one of the few defences left in (partial) democratic systems against powerful vested interests. In the EU the regulation of transgenics is undertaken in the interests of 'consumer choice' and of 'farmer choice'.

However, regulation issues raise questions about the potential for identifying and valuing public goods, such as clean air, water and food and these goods are not simply a matter of consumer or farmer preference.[72] Policy makers will claim that regulation is based on best evidence, on cost-benefit analysis and to a lesser extent on risk–benefit analysis. However, where the evidence is disputed it is necessary to be more explicit about the assumptions that lie behind regulatory decision-making processes.

Transgenic organisms have been subject to regulation from the outset, with early research work and biosafety checks taking place in a laboratory setting. Additional regulatory and safety issues arise in the context of releasing organisms into the wider environment. Transgenics are regulated in two distinct areas: first in relation to human welfare, as food organisms for human consumption or as health products for medical use, and second in the interests of environmental protection, in relation to ecosystem and animal health.

---

72   It is not the intention of this study to trace the development of the concept of the 'public good' in environmental ethics and politics, or indeed to enumerate those goods which might be considered 'public'. What is significant for this study is that the development of the term and its increasing use represents a move away from economistic oversimplifications in policy formation.

Substantial equivalence in food safety

The regulation of transgenic food, as regards human health, has been largely based on the concept of 'substantial equivalence'. The concept of substantial equivalence has come in for some scrutiny and there is no broad consensus that it is up to the task of assuring food safety. Using this concept, transgenic foods are classified according to their general similarity or dissimilarity to known traditional foods with a history of safe use. If deemed 'substantially equivalent' (as opposed to not substantially equivalent or not equivalent in a specific well-defined characteristic) foods are passed for human consumption.[73] There is no consensus that substantial equivalence warrants being classified as an 'evidence-based' approach and for some it serves to act as a barrier to further research into the possible risks of eating GM food.[74] Its 'official definition' is loose and vague, yet it remains at the hub of food safety regulation. It is no substitute for a practical approach that would actively investigate the safety, toxicology and implications for public health of transgenics.[75] Furthermore, the concept of 'substantial equivalence' will not suffice in the development of 'second' and 'third' generation organisms produced to manufacture particular pharmaceuticals and industrial products. The aim with such transgenic modification is to radically alter organisms in order to produce novel or increased quantities of secondary metabolites. Here the quantification of the quality, and the purity of secondary metabolites as products, will be the central aim of gene manipulation. In that case second-generation crops, 'which involve output modifications

---

73 'This approach is based on principles developed through international expert consultations carried out by the World Health Organisation and the Food and Agriculture Organisation of the United Nations as well as the Organisation for Economic Co-operation and Development. Currently this strategy is applied by regulatory agencies in the European Union, Australia/New Zealand, Japan and the United States'. Silva and Buchanan, 'Regulatory Considerations for Transgenic Animal Health and Food Safety Assessment; a Canadian Perspective', 188.
74 Millstone, Brunner, and Mayer, 'Beyond "Substantial Equivalence"', 525.
75 Ibid., 526.

(traits with health and nutritional benefits), will likely only be viable if their purity or quality can be assured'.[76] They would not be, in that case, substantially equivalent to the original crop.

Triggers for regulation

It is also significant that the 'triggers' for regulation vary with jurisdiction. In Canada the 'trigger' is novelty itself, all new organisms (transgenic or not) are subject to regulatory scrutiny. In the USA regulation is predicated only on identified unwanted outcomes of introducing a novel organism. In the EU the regulatory 'trigger' is the technique by which organisms are produced, with organisms produced by recombinant DNA techniques coming in for specific scrutiny.[77] In the EU, for example, the trigger for labelling transgenic food is the presence of protein or DNA from genetic modification. In January 2000, the European Commission adopted a regulation setting a threshold limit for the labelling of food products that contain genetically modified material, including inadvertent contamination.[78]

International perspectives on precaution and regulation

Two other concepts distinguish European regulation from that of the USA and Canada; the notion of precaution and the issue of case-by-case analysis. Europe has incorporated the concept of precaution into its legislation and the principle guides the Directives that form the legislative framework for the regulation of biotechnology.[79] The adoption of the precautionary principle in Europe has implications for the research and development of transgenics. The European approach to precaution

---

[76] S. Smyth, G. Khachatourians, and W. Phillips, 'Liabilities and Economics of Transgenic Crops', *Nature Biotechnology* 20 (2002), 537.
[77] Cf. J. Kinderlerer, 'Why Regulate and How?', *Environmental Biosafety Research* 2 (2003), 52–53.
[78] This does not apply to composite meals in restaurants for example.
[79] Kinderlerer, 'Why Regulate and How?' 53.

is interpreted as putting the onus on researchers and technologists to demonstrate biosafety.

Internationally, however, the precautionary principle is a contested concept. It is subject to more than one interpretation. In Europe it has been interpreted to err in favour of a case-by-case approach based on scientific evidence, albeit alongside the recognition of trade considerations. However, for Canadian and North American researchers, precaution, as interpreted by the EU, simply means inhibiting innovation in this area. The claim that case-by-case analysis does equate to a more strictly 'science-based' approach is likewise contested. The case-by-case approach is heavily criticised for imposing intolerable costs on the decision-making process and not just in monetary terms but also in 'philosophical' terms.[80] In the interests of 'prudence', Paul Thompson, a philosopher and ethicist, suggests that 'stacking the cost and benefits of each biotechnology product against one another is less reasonable than it appears'.[81] He suggests that operators must also make use of the 'cognitive filters' normally relied upon in decision-making processes. The question, for Thompson, increasingly becomes one of trust in decision-making institutions, both scientific and governmental. However that does not mean that the problems with this new technology can be reduced to a question of pessimism or of optimism, as Thompson would have it.[82] If an 'evidence-based' approach is truly a scientific goal then an 'optimistic' trust in just public institutions is not the only criterion for trustworthiness. Case-by-case analysis may constitute the 'state of the art', the best 'scientific' measure of the performance of transgenic organisms. Economic interests may conflict with it but do not therefore negate its scientific validity.

In many countries there are no regulations for transgenic organisms in place or they are in rudimentary form only: in Eastern Europe, the Far East (no reference is made to the Islamic Middle East), Africa, and Latin America, the Caribbean and even in China, where public institu-

---

80 Cf. Thompson, *Food Biotechnology*, 20–21.
81 Ibid., 20.
82 Ibid., 9–13.

tions dominate research and development.[83] In China, state-run research programmes in transgenics claim to operate on a case-by-case approach. In the USA, Canada and the EU, even given that an organism becomes subject to regulation, there is no agreed 'evidence-based' strategy for assessing the unintended and negative outcomes of introducing many of the organisms under development. It remains unclear whether public demands to develop regulatory structures for the management of transgenic releases imply the development of comprehensive risk assessment procedures post-release, or merely a presumption in favour of this technology that seeks to free-up regulations in the interests of business.

With no internationally agreed evidence-based approach to the regulation of transgenic technology, countries in the 'developing world' look to the dominant models of the USA, Canada and the EU in search of an approach that addresses the demands of their own contexts. It seems clear that the development of agriculture, in some African countries for example, may not necessarily be tied to more technology, particularly where that leads to greater dependency. The issue, for indigenous agriculture, whether in Ireland or Kenya, is one of appropriate technology transfer in agriculture and industry.[84]

There have been some attempts to develop international standards in the regulation of transgenics (although these are often subject to competing, often overriding international concerns, not least questions of trade). The Cartagena Protocol[85] is one such regulatory tool. It requires its signatories to have in place a regulatory system for handling *living* modified organisms which could impact on conservation and the sustainable use

---

83 Cf. Proceedings of the 6th International Symposium on the Biosafety of Genetically Modified Organisms, University of Saskatchewan, 2000.
84 Cf. M. Kandawa-Schulz, 'Biosafety Regulations, Releases and Research Activities in Selected African Countries' (paper presented at the Proceedings of the 6th International Symposium on the Biosafety of Genetically Modified Organisms, University of Saskatchewan, 2000), 6–15.
85 The Protocol, adopted in Montreal, Canada was called after a Columbian conference in 1999.

of biodiversity.[86] The protocol is concerned only with the safe transfer, handling and use of living modified organisms. It does not address food safety issues, the segregation of bulk commodities, or consumer labelling. It requires 'an advanced information agreement' on the movement, transnationally, of transgenic living organisms like seed, fish etc. It has been ratified by more than 50 states and came into effect in 2003. The US is not a signatory although it was involved in the drafting of the protocol. Nevertheless it is considered by the EU as 'an important international instrument reconciling environment, trade and development', in that it offers a framework for the beneficial development of biotechnology world-wide.[87]

The protocol has been criticised for being at its heart yet another trade agreement and not a serious attempt to deal with the potential impact of transgenic organism. This criticism and others illustrate the considerable problems for regulators in ethical terms. Although it may seem that the system of regulation for biotechnology has been proactive[88] it is clear that any decision to regulate biotechnology is a *de facto* decision in favour of using it. 'The resulting regulatory frameworks reflect the principal agreement to use the tools provided'.[89] Regulation cannot however of itself deliver on food security and environmental biosafety. As de Kathen points out, 'technical guidelines are prominent products but only part of a biosafety system that also needs people, a review proc-

---

86 Kinderlerer, 'Why Regulate and How?' 51–56.
87 J. Kioussi, 'The Evolution of the EU Regulatory System for GMOs' (paper presented at the Proceedings of the 6th International Symposium on the Biosafety of Genetically Modified Organisms, University of Saskatchewan, 2000), 29.
88 J. Kinderlerer, 'Transfer of Genetic Modification Technology into the Third World: Biosafety and Developing Countries' (paper presented at the Proceedings of the 6th International Symposium on the Biosafety of Genetically Modified Organisms, University of Saskatchewan, 2000), 71.
89 A. de Kathen, 'Biosafety Capacity Development and the Cartagena Protocol' (paper presented at the Proceedings of the 6th International Symposium on the Biosafety of Genetically Modified Organisms, University of Saskatchewan, 2000), 79.

ess and a feedback mechanism for improvement'.[90] He suggests that the Cartagena Protocol acknowledges that biotechnology and biosafety are not just technical issues for science and economics, but have significant social impact. If biotechnology is to be worthwhile, for developing countries for example, then it needs to be regulated to their advantage, in terms of human health, food security and the environment, as well as trade. He suggests that it also needs to be financed to their advantage. Regulation should include some mechanisms by which goal-directed improvements might be financed, either by a sharing of the subsidies currently provided in agriculture in the West or by the payment of significant royalties for access to native 'genetic' resources.

> To improve the situation for developing countries, R&D activities in biotechnology have to change in focus and volume – HIV, malaria, and the generation of robust and pest-resistant crops (cassava, sweet potato, millet, etc.) are prime targets for research. Financial provisions may come from an equitable share on subsidies currently provided to the industry and royalties on crop seeds or medicinal plants.[91]

Such proposals may need further ethical investigation since they imply profound changes for agriculture and industry in Western democracies also.[92] Biosafety protocols, like Cartagena, may make a contribution to biosafety but only where the regulatory process is seen as continuous, interdisciplinary and integrative in nature.

> By perceiving biosafety capacity development as a continuous and interdisciplinary process and by reflecting the scope and objectives of the Cartagena Protocol, assistance in elaborating and implementing biosafety frameworks will need to take into account policy development, legislation, assessment procedures, and institutional and human capacity as equally important elements of an integrated approach.[93]

90 Ibid., 84.
91 Ibid., 86.
92 The termination of the sugar beet industry in Ireland (effective since spring 2006) was a direct result of the planned removal of EU subsidies, in the interests of trade with sugar-producing countries in the developing world.
93 de Kathen, 'Biosafety Capacity Development and the Cartagena Protocol', 86.

Regulation reflects, only in a partial and limited way, the scientific, economic, social and ethical evaluation of transgenic technology. The demand that regulation be 'evidence-based' has the effect of restricting it to post-release or post-introduction evaluations only. It is clear that the scientific assessment and regulation of this technology is already framed by a set of assumptions that need to be made more explicit for regulatory purposes. In the next section I will examine the scope and limitations of cost-benefit analysis and suggest that government policy must be based on more than economistic measures. It cannot neglect the cultural and institutional framing of the decision-making process.

The role of cost–benefit analysis

In this section I would like to focus on economic strategies in environmental management. Economics could be described as the 'science of choice' and as such economists have attempted to find methods to account for the value of non-traded environmental and public goods. They do this in order to improve the 'provision' of these goods and to take account of their depletion or degradation. They seek to broaden the scope of cost–benefit analysis (CBA) to include the value of environmental goods previously considered 'external' to the considerations of the market. Although well-intentioned, these attempts have led some economists, not to the satisfaction of ever more accurate accounting, but to an acknowledgement of the limits of the neo-classical paradigm and economistic thinking in general. Ironically, although some economists themselves now dispute the internal consistencies of CBA, it has become an important strategy in policy making with governments and international bodies world-wide.

It should be noted that although cost–benefit analysis, risk–benefit analysis (RBA) and risk assessment are related terms, they are not interchangeable in all contexts. Cost–benefit analysis is generally the preferred term for classical economists. Environmental economists tend to replace cost–benefit analysis with risk–benefit analysis, while environmental managers and scientists tend to talk in terms of risk assessment. The latter two are concerned more with the quantification of critical loads and the carrying capacity of ecosystems rather than trade-offs. They consider the limits put on economic activity, rather than weighing up benefits and trading off harms. There is some dispute therefore among economists as to how economic imperatives should be factored in in environmental policy. This has implications for the discussion of recombinant DNA technologies in food production.

In scientific perspective risk–benefit and cost–benefit analysis are increasingly expected to carry the requirements of objectivity, utility and transparency in policy-making in the area of biosafety. CBA already enjoys a central place in public economic policy in many spheres and has gained ground in environmental economics. This is generally viewed as a positive step. After all it elevates environmental concerns by translating them 'into the grown-up language of economics' making environmental policies an important, quantifiable governmental concern.[94] The extension of neo-classical economic theory to include the value of 'natural capital' has, on the positive side, prevented policy makers from utterly undervaluing non-marketed environmental resources and outputs. So it serves to rein in the drive to undifferentiated economic growth that is perceived as the major threat to the environment.

CBA is an approach in economics[95] that balances the monetary cost and benefits of a particular course of action. It 'is founded on the normative political theory that public policy decisions should be made on the basis of aggregating private preference based choices'.[96] Environmental economics, as a subdiscipline of economics, broadens the application of CBA and attempts to ameliorate the evident failure of actual markets to efficiently count and cost the value of environmental resources. It is this failure, the failure to account for non-traded goods, that is assumed to

---

94 Robin Grove-White, 'The Environmental "Valuation" Controversy: Observations on Its Recent History and Significance', in *Valuing Nature? Ethics, Economics and Environment*, ed. John Forster (London: Routledge, 1997), 23.
95 The study of choice under conditions of scarcity. John Foster, ed., *Valuing Nature? Economics, Ethics and the Environment* (London: Routledge, 1997), 4.
96 Michael Jacobs, 'Environment and Creative Value', in *Valuing Nature? Economics, Ethics and Environment*, ed. John Foster (London: Routledge, 1997), 215.

contribute to ongoing environmental degradation. CBA can be broadened to include the value of non-traded goods through 'contingent-valuation methods' or 'willingness-to-pay methods' which make up the shortfall in accounting for hidden costs or benefits (referred to as externalities), in relation to natural capital.[97] The calculation of 'contingent valuations' and 'existence values' are referred to as surrogate valuation methods.[98]

In the neo-classical model economic analysis aims to predict behaviour by observing the real market. Calculation is based on actual preferences expressed in real markets; the decision-making process is irrelevant to the economic outcome in that case. Economic theory so conceived, in theory at least, equates utility with the good.[99] In contrast, contingent valuation relies on the field of survey research which attempts to approximate preferences.[100] Some reservations have been expressed among economists about the reliability of survey techniques and the need to test predictions against actual behaviour. 'For many economists, the ultimate argument against contingent valuation is that it violates the habitual commitment of the profession to revealed preferences'.[101] Certainly for some economists contingent valuation represents a paradigm shift in economic thinking; it attempts to be precise about a non-market valuation.[102] For that reason it cannot account for non-market values in all cases. Indeed, it is suggested, these techniques may represent misplaced or reduced precision.

---

97 Russell Keat, 'Values and Preferences in Neo-Classical Environmental Economics', in *Valuing Nature? Economics, Ethics and the Environment*, ed. John Foster (London: Routledge, 1997), 32.
98 Grove-White, 'The Environmental "Valuation" Controversy: Observations on Its Recent History and Significance', 21.
99 Cf. Jonathan Aldred, 'Existence Value, Moral Commitments and in-Kind Valuation', in *Valuing Nature? Economics, Ethics and the Environment*, ed. John Foster (London: Routledge, 1997), 155–69.
100 W.M. Hanemann, 'Valuing the Environment through Contingent Valuation', *Journal of Economic Perspectives* 8, no. 4 (1994), 21.
101 Although revealed preferences are not themselves fool proof and involve extrapolation. Ibid., 36.
102 Cf. Ibid., 37–38.

Environmental policy-making is not necessarily wholly dependent on neo-classical economics, of course. Spash differentiates between the economic approach and what he calls the 'environmental management approach' usually associated with the natural sciences.[103] These two, he suggests are methodologically disunited, one being concerned with trade-offs, the other with limits. 'A conflict arises because economics is primarily concerned with trade-offs, while the natural science approach to environmental management has been concerned primarily with limits (e.g. sustainable fishing yields, carrying capacity, pollution thresholds)'.[104] The natural sciences play a role in that they put constraints on economic policy. The environmental management approach doesn't attempt to measure benefits but concentrates instead on the cost-effectiveness of controlling human activity which may be damaging to the environment.[105]

Although not involved necessarily in trade-offs, the management approach can also suffer the same problems as the cost–benefit approach; it is open to the same criticism in that 'the appeal to facts based on empiricism creates a false sense of freedom from moral and other value judgements.[106] It is unclear to what extent economics can play a role in environmental decision making but, it could be claimed that there is clearly a 'false sense of scientific objectivity' in current models.

The internal economic inconsistencies of contingent valuation techniques and neo-classical analysis are not all that is problematic conceptually. In risk–benefit analysis, risk is interpreted to have a universal and quantifiable meaning, which can be traded against benefits, similarly defined. Both, it is assumed, can be translated into monetary units outside of the contexts in which they arise.[107] Human lives (valued at

---

103 C. Spash, 'Environmental Management without Environmental Valuation', in *Valuing Nature? Economics, Ethics and the Environment*, ed. John Foster (London: Routledge, 1997), 170–85.
104 Ibid., 170.
105 Ibid., 178.
106 Ibid., 183.
107 Brian Wynne, 'Methodology and Institutions: Values as Seen from the Field', in *Valuing Nature? Economics, Ethics and the Environment*, ed. John Foster (London: Routledge, 1997), 140.

€1.3 M), public, social and economic goods and non-renewable resources are counted alongside commodities like coffee and orange juice (which carry their own costs in human lives, goods and resources).

Yet, however much economic comparisons are useful in weighing up possibilities, they carry with them an attendant danger. Most notably surrogate valuation methods require that a price be put on goods not readily subject to monetary evaluation. Costs and benefits may be incommensurable. Contingent valuation, therefore, is not theory neutral but in fact encourages a consumer approach.[108] The issue is not whether values represented this way may have utility, but whether economic valuation becomes 'normative' for every value; whether it becomes a given to which we must conform rather than being a tool in the service of policy makers. Thompson suggests, in defence of RBA, that it is at least a 'way of organising information that is relatively objective and allows everyone to see where key philosophical decisions have been made.'[109] Yet the institutionalisation of cost– and risk–benefit analysis also has the effect of already framing its conclusions, of tacitly promoting a particular form.[110] If the elaboration of methodological questions obscures institutional ones then there is a good case to be made, not for it, but for institutional reform, as Hodgson suggests. A shift to an institutional approach in ethics can serve as a part corrective to the hegemony of the neo-classical approach; which pictures the public good, simplistically, as the sum of individual preferences.

> Institutional economics promotes a quite different view of the nature of the economy. It involves a distinctive kind of analysis, emphasising the cultural and institutional framing of decision and action. Instead of being regarded as mere constraints or ephemera, institutions, culture and values are regarded as the stuff of socio-economic life.[111]

---

108 Jacobs, 'Environment and Creative Value', 216.
109 Thompson, *Food Biotechnology*, 111.
110 Ibid., 144.
111 G. Hodgson, 'Economics, Environmental Policy and the Transcendence of Utilitarianism', in *Valuing Nature? Economics, Ethics and the Environment*, ed. John Foster (London: Routledge, 1997), 61.

The attempt to measure the value of a 'public good', such as education or clean water or air, has led steadily to a conclusion that the public good is more than the sum total of individual preferences. The public good includes consideration of preference but also of externalities, of ethical principles and of the common good.[112] The environment is a public good because it is more than simply of interest to the individual. It is also more than the sum total of all individual valuations, now and in the future. This is an aspect of the common good, which will be discussed in chapter two. The role of institutions, as regards the public good, is neither to aggregate individual preferences to give an overall societal preference, nor to pronounce that an optimum is reached when benefits exceed costs.[113] Surrogate valuations that modify the cost–benefit approach do not overcome its shortcomings: 'public policy decisions should be made on the basis of some conception of the public good, which is logically separate from the aggregations of individual private benefits and costs.'[114]

Jacobs suggests that institutional change is called for, among environmental managers and in professional cultures; in particular for cases where there are no 'rational measures' that can easily replace practical judgement.[115] The aim would be to create a new framework for small-scale management decisions which would remain open to public scrutiny, with rights of participation and appeal. Institutions, after all, not only reveal, but actively construct, values. As Jacobs suggests public goods are less a commodity and more a 'context of life'. Economic techniques have utility, but do not replace the consideration of alternatives, the need for the integration of differences nor the role of a discursive framework.

112   Jacobs, 'Environment and Creative Value', 215.
113   Ibid., 212–14.
114   Ibid., 215.
115   Ibid., 224–25.

Summary and evaluation

Environmental valuation using neo-classical cost–benefit analysis is limited in scope and this is the case even with the development of surrogate valuation techniques. Yet cost–benefit analysis and its variants, risk–benefit analysis and risk assessment, continue to be the strategies most relied on in the development of policy around new technologies; in relation to the deliberate release of transgenic organisms for human consumption and into the environment for example. It is unclear what role such analysis should play in the decision-making process. Their 'utility' and 'objectivity' cannot be uncritically assumed, even as they may serve to provide one scale by which to judge the environmental consequences of human activity. There is some consensus that institutions need to better understand their role; 'which is not (the methodological) one of calculating and communicating the right decision'.[116] Rather it is to continue the participative process of building just institutions. In relation to food biotechnology Thompson concludes that 'trust building communication remains the most precarious and least fulfilled ethical responsibility for the proponents and practitioners of food biotechnology'.[117] In other words the debate over transgenic technology not only affects public trust in science but also in mandated governmental and social institutions that mediate between the private and the public good. Instead of resting with a critique of the shortfalls in the classical theory of exchange value, environmental economists have identified the need for institutional reform and the reestablishment of a notion of 'practical judgement' and 'public goods' as alternatives to economistic thinking. The notion of 'sustainability' has some leverage with both environmental managers and productionists. As such it will be discussed next as a 'concept of convergence' across several disciplines – environmental science, economics and ethics.

---

116   Wynne, 'Methodology and Institutions: Values as Seen from the Field', 136.
117   Thompson tends to reduce the discussion to a question of a choice between optimism and pessimism (his choice being optimism) in dealing with this new technology. Thompson, *Food Biotechnology*, 17.

Sustainability: normative and descriptive aspects

Sustainability can loosely be defined in environmental ethics as the ability of non-renewable resources to persist over time. It has the potential to reconnect polarised factions in disputes over new technologies in agriculture and to create a space for the development of innovative and agreed procedures in the area of environmental protection. For Thompson, it offers an alternative to the overplayed dichotomy that pitches ecosystem preservation against humanity:

> Ecocentric language can specify duties to preserve or respect nature, but presents little guidance with respect to environments in which agriculture has already replaced the natural order. Anthropocentric language can specify duties to conserve resources for future use, but fails to build any concept of ecosystem integrity into the basis of moral obligation.[118]

Thompson argues that the concept of sustainability has some ethical promise for integrating these seemingly divergent viewpoints.[119] In environmental ethics it has both heuristic and integrative potential. Although frequently subject to redefinition and revision, what is significant from an ethical perspective, is that it has both a descriptive and a prescriptive aspect. As a *descriptive* concept it needs to be firmly based in the sciences and in economics. As a *normative* principle it requires a social and ethical foundation. It mediates between production and protection in agriculture and it is inclusive of the concept of intergenerational justice. It can therefore also be compatible with just development, current and future, as a social and ethical goal.

---

118   Thompson, *The Spirit of the Soil*, 148. Paul Thompson was formerly the Director of the Centre for Biotechnology Policy and Ethics and Professor of Philosophy and Agricultural Economics at Texas A&M University and is now the Joyce and Edward E. Brewer Professor of Applied Ethics at Purdue University, Indiana, USA.
119   Ibid.

The concept of sustainability has gone some way to integrate the aims of productionists and environmentalists in the context of the unwanted effects of agricultural production and technology on the environment. As a concept in environmental ethics, it is the bearer of at least two essential ideas. First it registers the dangers of the permanent growth of agricultural and industrial productivity through the destruction of non-renewable resources. Second it recognises the rights and needs of future generations, as well as the present generation, in relation to the same resources. Admittedly the emphasis on 'intergenerational justice' has sometimes been overstated to the point of downplaying questions of justice for present generations.[120] The stress on the rights of future generations, over the current generation, has in turn lead to contrasting corrective moves. Corrective approaches suggest that the present generation may need to 'spend' more than its fair share of natural capital, in order to secure more sustainable production methods and resources for future generations.

It is possible to define sustainability such that it makes no philosophical advance on utilitarianism or cost–benefit analysis.[121] The promotion of self-regulating systems, in the biological sense, may have utility but need not imply an ethical goal.[122] However, scientific and economic models

120 The 'one child policy' in China serves to illustrate this. In his comparative analysis of population control measures, Amartya Sen concludes that compulsion by the state in China not only leads to unintended and undesirable problems such as the increased violent loss of rights, increased 'infant mortality' rates, the fatal neglect of female children and even an unaltered attitude to reducing fertility, there is no evidence to support the claim that coercion can deliver any extra population control reductions over and above education and improved opportunities, in particular for women. He makes a case for voluntarism, as opposed to coercion, based on education for women, as the instrument of lasting and even faster population control. It is noteworthy that his comparison is not made with the affluent West or even with the 'rich' in poorer countries but with Kerala, a poor Indian state with a relatively good education system. Amartya Sen, *Development as Freedom* (Oxford: University Press, 1999), 204–23.
121 Thompson, *The Spirit of the Soil*, 149.
122 Sustainability has traditionally been equated with sustained yield in Irish forestry for example. In order to embrace all the goods and services of the forest sustained yield has been replaced with the broader concept of 'sustained management'. Cf.

of environmentally sustainable systems understate the social and ethical aspects of this concept. For Thompson these two aspects – the descriptive and the prescriptive – are often undifferentiated and confused. 'The system-describing and the goal-prescribing concepts of sustainability are now applied indiscriminately in alternative agriculture and development studies'.[123] From one perspective, sustainability acts as the post-modern equivalent of progress and the ideal goal to which all reform should be directed.[124] From the other it has little to do with ethics or morality.

For Thompson, these two aspects need to be made more explicit. The system-describing approach, he suggests, is most appropriate to the sciences. In food production and agriculture it is not possible to do without a 'judicious understanding of the natural and agricultural systems'.[125] What the natural sciences contribute to the analysis is a description of the possible outcomes open to policy makers alongside other goals in society and ethics.

What is implied is that there can be no 'neutral' assessment of sustainability. The goal of sustainability in any system depends on how that system is modelled.[126] Systems do not have 'natural' borders, rather these are constructed in relation to the problem that a system-analysis is expected to resolve.[127] The choice of borders and performance criteria

---

Edward P. Farrell, 'Managing Our Forests for the Future' (paper presented at the Annual Symposium of the Society of Irish Foresters, Newbridge, 1996).
123 Thompson, *The Spirit of the Soil*, 153.
124 Ibid., 172.
125 Ibid., 166.
126 As Thompson points out; 'Systems are analytically defined with borders and with internal interactions. A systems model takes a few characteristics of the internal elements, such as population, calorie consumption, or even profit to be significant, then determines the impact of systems interactions upon these characteristics, assuming a constant (or at least regular) exchange of input and output across systems borders. In standard systems analysis, performance criteria for these characteristics will be given as a range of permissible values, and the system will be said to have failed when the range is exceeded by either the upper or lower bound'. Ibid., 154.
127 Ibid., 156.

are open to dispute. Sustainability criteria, therefore, necessarily have social and ethical content.

Undoubtedly, in contrast to utilitarian models, it is possible to incorporate too many aspirational values into a system, making it impossible for a system to come out as 'sustainable'.[128] A system that ties global development to permanent economic growth is likely to be 'unsustainable'; collective human demands, in theory at least, could exceed the carrying capacity of the biosphere. Likewise a system that is not committed to distributive justice could be 'unsustainable', as regards non-renewable resources. Poor countries may not be in a position to forego development opportunities that consume natural resources, such as forests or biodiversity, without international co-operation to close the incentive gap.[129]

Sustainability has currency across several disciplines, not least science, economics, ethics and policy-making. It has the capacity to integrate the descriptive potential of the natural sciences and economics with the normative aspects of social and ethical analysis. Thompson's analysis as presented here indicates a clear role for the natural sciences; one of attempting to describe the environmental consequences of introducing new technologies, such as transgenics, laying bare the implicit assumptions made in doing so. There are also philosophical and ethical assumptions made in any description of a sustainable system and these also need to be made clear. Thompson suggests that it 'is less important to advocate sustainability as a political or ethical goal, than it is to make the philosophical assumptions in our understanding of natural and human systems explicit'.[130]

---

128   Ibid., 159.
129   Inge Kaul, Isabelle Grunberg, and Marc A. Stern, *Global Public Goods: International Cooperation in the 21st Century* (Oxford: Oxford University Press, 1999), xxiv.
130   Thompson, *The Spirit of the Soil*, 167.

Conclusions

I have described some of the developments in transgenic technology in the area of food production and some of the unintended but identifiable consequences of developing that technology. In scientific perspective, the problems associated with recombinant DNA techniques are described as empirical and consequential in nature. The assumption is, that because they are technical in nature, they are subject to amelioration through further technical breakthroughs. This assumption is not without merit. Environmental problems, particularly those that have had a long period of *ad hoc* development, may well be ameliorated by technical means. The Irish Environmental Protection Agency (EPA) currently funds research into technical solutions for environmental issues. However, the commitment to amelioration and to research ought to imply that we do not continue to knowingly create further technical problems. In addition, weighing the technical outcomes in a 'technological ethics', as Thompson tentatively suggests, could merely 'qualify the direction and application of biotechnology'.[131] It presumes that trade-offs can continue to be made, rather than limits set. It is not designed to be preventative, but merely reactive or indeed simply profitable.

Policy makers have developed strategies to deal with some of the unintended consequences of new technologies. The regulation of transgenic organisms is one such strategy, another is the application of an expanded cost–benefit analysis in the environmental sphere, to calculate the risks and the benefits of innovations. These strategies can be partial only, and as such, suffer from both internal conceptual inconsistencies and from empirical limitations. They do not substitute for the institutional value frameworks that are responsible for interpreting, evaluating and implementing policy.[132]

The limitations of regulatory and economic strategies to value the public and common good have also inspired new approaches in

---

131 Thompson, *Food Biotechnology*, 199.
132 Wynne, 'Methodology and Institutions: Values as Seen from the Field', 135–52.

environmental ethics and philosophy. Sustainability has served as a concept of convergence for productionists, economists and natural scientists alike. The suggestion is that agriculture needs an ethic of the environment and environmentalists need an ethic of production.[133] Sustainability has some potential in articulating these needs. In distinguishing the descriptive and the prescriptive element in the concept of sustainability, Thompson implies that the natural sciences make their most significant contribution in the descriptive mode.

Sustainability holds out the possibility that insights from science, economics, and ethics can be integrated in environmental ethics. The role of the sciences is to discover what levels of use and reuse of natural resources can continue indefinitely. The role of economics is to reflect and internalise these in choices about non-renewable resource use. The role of ethics is to secure the moral framework and values in which these disciplines serve the human person in society. Sustainability captures many concerns that are also at the same time theological: the moral and ethical significance of non-human nature as the context for created finite rational nature, the importance of the common good, and the value of non-human created reality.

The suggestion that transgenic technology will help to secure a more 'sustainable' agriculture is a broad-reaching claim and one that needs further testing. It has not been convincingly substantiated from a scientific and economic perspective. The significant ethical task, for the purposes of this study, is to now investigate how sustainability, as a descriptive enterprise, fits with other goals in society and ethics. How for example it might fit with a commitment to distributive justice and to development, both of which are predicated on the prior recognition of the dignity of each human person. In order to engage in this task it is first necessary to expand on the goal-describing aspect of sustainability, to examine the ethical and moral framework in which a descriptive theory of sustainability could be articulated. It is to the development of such an ethical framework that I now turn.

---

133   Thompson, *The Spirit of the Soil*, 15.

CHAPTER 2

An Autonomy Perspective in Theological Ethics

In chapter one the scientific and economic aspects of transgenic technology were examined, both in relation to human health and in relation to wider environmental issues. Institutional, regulatory and economic strategies aimed at delivering 'sustainability' suggest that agriculture needs an ethic of the environment and that environmentalists need an ethic of production. Science and economics contribute to the descriptive aspects of sustainability, outlining the possibilities and choices that are potentially open to policy makers and regulators. In contrast the normative aspects of sustainability outline the frameworks within which these choices are articulated, realised and institutionalised. Therefore I will first make explicit the normative presuppositions and fault lines along which I defend and articulate an ethical framework for agricultural production and environmental ethics from a theological perspective. The position I advocate – the autonomy position – draws on theological concepts that are also defended, although differently nuanced, in other theological approaches. Here I will present three theological positions in ethics[1] that serve both as sources and as contrasting approaches against which to articulate my position – autonomy in a Christian context.

Theological positions in ethics exhibit many parallels, strengths and weaknesses, with predominant, if much disputed, moral philosophies. Several possible starting points share themes and resources with each other while diverging in content and conclusions.[2] It is a characteristic

---

1   The four in total presented here do not exhaust the field of theological ethics. Investigations could be conducted from other points of departure; from a postmodern critique of power or a feminist critique of patriarchy for example.
2   These subdivisions did not originate in, but correspond in some important respects, to those discussed (with a different emphasis and order) by Porter in *The Recovery of*

of ethics that it is constantly reintegrating its sources and arguments in relation to issues arising in lived contexts. Three theological 'sources' are outlined here: divine command or evangelical ethics,[3] Christian communitarian or ecclesial ethics[4] and natural law.[5] I will argue that 'autonomy in a Christian context' is best placed to integrate the insights from theology, philosophy and science in developing an ethics of sustainable production in agriculture.[6]

The autonomy approach that I outline and defend, draws on and integrates the theological concepts and critiques articulated in other ethical positions in theology. It is open to the insights of Divine Command morality which insists that scriptural revelation critically shapes and 'constrains' Christian ethics. It shares with it the concern to elaborate on how the Biblical text informs ethics and influences the Christian evaluation and legitimation of cultural moral and ethical practices. Both divine command morality and autonomy make explicit the relativising aspect of the Christian faith understanding that the biblical text refers to more than the liberation of the human person in morality and ethics. However, the autonomy approach methodologically separates the question of meaning and faith, and the possibility of salvation and fellowship with God, from the question of moral obligation. In divine command morality our moral obligation is theologically justified; in contrast in the autonomy position it is justified philosophically. Perhaps because of this, the autonomy approach is often in a strong position to make explicit, and to acknowledge, its debt to the intellectual resources of non-theological disciplines.

Virtue: a theocentric return to nature; the search for the virtuous community; the focus on human goods and actions in revised natural law since the Second Vatican Council; and lastly Christian love as equal regard, self-sacrifice and mutuality.

3   I make reference in particular to the work of O. O'Donovan.
4   I make reference to the work of A. MacIntyre and S. Hauerwas in his use of MacIntyre, and as discussed by D. Fergusson.
5   With reference to R. Gula, L. Sowle Cahill, Jean Porter and A. Lisska.
6   Taking D. Mieth and M. Junker-Kenny as current representatives of this position in theological ethics, initially developed by A. Auer, J. Fuchs and I. Boeckle.

In its theological reconstruction it also defends, alongside communitarian approaches, human sociability. Kant conceived of the human person as an 'end in him/herself' in a 'Kingdom of Ends'. Indeed the freedom of the other is a condition of our own moral freedom and not external to it. In that the autonomy approach is in keeping with the communitarian concern to develop an ethic of tradition. It acknowledges the role of community and sociability, which confirms and sustains the human person, allowing them, in solidarity, to live up to their moral obligation and creating the conditions which enliven and reinforce Christian value systems. However, unlike some radical forms of communitarianism, the autonomy approach also keeps the role of human freedom to the fore, as the condition for the possibility of morality in the first instance. It does not subsume individual freedom to community concerns. It is relevant, for example, to this study that an autonomy approach would not advocate coercion in human reproduction, in the interests of environmental preservation. It calls for ways to reach environmental goals that are in keeping with the dignity of the human person.

Each of these positions will be described and weighed in relation to four guiding themes: Christian distinctiveness and the place of scripture in ethics, the relationship between deontology and teleology, a commitment to a shared or public moral discourse, and lastly their implications for the relationship between the human person and non-human nature. Each position articulates an important theological concern as a different starting point, the divine authority of revelation in Christian ethics, the normative role of the Christian community for individuals and societies, the moral significance of human rational nature and lastly the transcendental status of human freedom in Christian ethics. Each implies a particular role for the deontological and teleological elements in any theological ethics. Morality in deontological mode investigates the ground of obligation prior to the outcome of actions. Deontological theories are contrasted with teleological theories which give priority to the good over the right, with the good being defined as the end or purpose of human action. Ethical positions in theology tend to avoid any sharp dichotomy between these two but, depending on their starting positions, weigh them differently.

These four starting points also assign a different weight to what counts as Christian, and to the place of scripture, in ethical discourse. In divine command ethics the Christian *proprium* is sought and found in scripture. In communitarianism, at least in its nascent form, the emphasis is firmly teleological, focusing on the goals of the good life and the formation of community. Natural law is grounded in Thomas Aquinas's reconstruction of Aristotelian teleology, although it defends a prior deontological force, a force that is grounded in a theological assumption. Its strengths lie in its long-standing teleological outlook, from which developed, in Catholic social teaching, the concepts of the common good and its corollary subsidiarity, both of which have significant appeal for an environmental ethics. In the autonomy position, however in contrast to the above three, ethical obligations are grounded not theologically but philosophically in human freedom. The theological *proprium* is found in the Christian context of this ethics of freedom, in which our hopes, our obligations and our task are articulated.

The combination of these elements influences how Christian moral insights are gathered in ethics but also how they are translated outside of the Christian faith community. They also impact on any reconstruction of nature, human and non-human, that may inform an environmental ethic. The objective is to discover what resources these approaches provide to address the need for a philosophy and theology of nature and of production, something that is not explicit, nor often even in evidence (except in often narrow utilitarian terms), in the technical application of the natural sciences. They contribute to a reconstruction of a philosophy and theology of nature and agriculture that does justice to the theological understanding of the world as created, to philosophical insights into the relationship between the human bodily realm and the wider ecological continuum, and to scientific insights into the natural world.

The theologians that are presented in this study have all addressed the question of the moral significance of the non-human creation in their ethical deliberation. Oliver O'Donovan revisits the question of the subordination of creation to redemption, that is taken as a defining characteristic of the divine command position. It is his rapprochement of creation and redemption that is the touchstone for the work of Michael

Northcott, whose particular approach to an environmental 'natural law' is critiqued in chapter four. Northcott's *The Environment and Christian Ethics* has been endorsed and promoted by several theologians and features in many current environmental theologies.[7]

Although no specifically communitarian approach has thoroughly dealt with environmental questions, some communitarian concerns, particularly in relation to the recovery of a Christian virtue ethics has been developed by Celia Deane-Drummond. Her virtue approach to an ethic of nature is critiqued in chapter four. Natural law approaches make a contribution to an ethic of the environment through their continuing emphasis on the connection between rationality (Jean Porter) and embodiment (Lisa Sowle Cahill). They have made significant contributions, in relation to environmental questions, in the area of institutional ethics, specifically through the principles of the common good and subsidiarity.

Dietmar Mieth and Maureen Junker-Kenny represent the Christian autonomy position which relocates the Kantian understanding of the right in a Christian context and argues for a rapprochement of the right and the good in that context.[8] Mieth specifically expands the categorical imperative to take account of the natural world as the true context for finite human freedom. His work points to the shortcomings of institutional approaches to ethics, particularly in the area of the regulation of new genetic technologies that need to be addressed. The churches are well placed to contribute to the debate. The concepts of the common good and of subsidiarity already show the potential in the value systems of church communities. They point to the need to develop more adequate frameworks, in the public realm, to interpret and evaluate technologies before implementing policy. The process of decision-making, as we have seen, has to be more than the sum of individual preferences, it must take

---

[7] 'One of the finest retrievals of natural law ethics, and particularly of Thomas Aquinas's version, is provided by Michael S. Northcott, in the context of working out a theological ethics adequate to respect our physical environment.' Fergus Kerr, *After Aquinas* (Oxford: Blackwell, 2002), 102.

[8] This approach therefore is not a philosophical deontology simply overlaid with a theological teleology.

account of ethical principles and externalities, and remain open to scrutiny in a discursive framework. Statutory bodies, such as the Environmental Protection Agency, or the Irish Council for Bioethics, need to be capable of considering alternatives, of integrating difference and of maintaining a discursive framework.

Alongside the emphasis on the institutional framing of values, the specific contribution the autonomy approach makes in environmental ethics is to recognise the full dignity of the person, showing that 'nature' is the befitting space for human freedom. It is precisely because the autonomy approach does not collapse the distinction between the human person and the non-human creation – and would argue for a hierarchy that put human needs first in times of conflict – that it is well placed to argue for human responsibility and obligation towards the non-human creation.

Divine command ethics or evangelical ethics

Divine command ethics has the advantage of being strongly biblically based, emphasising biblical and narrative theology in it interpretation of ethical issues. It is this emphasis that gives it its particularly Christian credentials. It aims to critically re-appropriate insights from the tradition that have often been expressed in the language of stories. In this way divine command morality contributes to a discussion of values in the public realm and in current contexts.

However, there is much internal Christian opposition to moral systems which claim as their only source of obligation the revealed commands of God.[9] The reliance of divine command morality on revelation as the one source of obligation, understood as the Word of God

---

9   Paul Rooney, 'Divine Commands, Natural Law and Aquinas', *The Scottish Journal of Religious Studies* xvi, no. 2 (1995), 117.

in Scripture, is both its major strength and its greatest drawback. All Christian ethical approaches consider the normative place of scripture in ethics but also give considerable weight to other sources, to tradition and to the prescriptive and descriptive aspects of human reason. In contrast divine command morality is characterised as relying on a direct appeal to Scripture as its point of departure, downplaying these other sources. Human reason and freedom, being created but fallen, are subordinated to the revealed Word.

Divine command morality, since it seems to require relinquishing any claim to autonomy in morality, is also often considered to be simply heteronomy.[10] It is not infrequently cited as one of the poles in the ethical discussion of Plato's *Euthyphro's Dilemma*, which is most simply formulated as follows – does God command what is good? Or is the good what God commands?[11] However, although for Barthian ethics our moral behaviour is as nothing compared with the love and mercy of God who 'loves us first', this cannot be interpreted to mean that faith allows the suspension of the moral law. God's 'command' cannot be dissociated from the right or the good, as understood, however imperfectly, by the human person. O'Donovan's 'realist principle' can be read as a reinvestment in the autonomy of the created world and in human freedom. It allows for the possibility that the universe as we perceive it corresponds, albeit not without remainder, with human rationality, freedom and morality and that the human person 'fits' there.[12] To see how divine command morality contributes to Christian ethics we must look more closely at the

---

10 Cf. P. Quinn, *Divine Commands and Moral Requirements* (Oxford: Clarendon, 1978).
11 This is a question also put to moral philosophy by Kant but with a radically different transcendental response. A divine command moralist could see these two poles as the two sides of the one coin, as Rooney does, in that it is possible to conclude from both that the existence of an objective moral law (in one derived from *Divine Fiat* and the other from the *Categorical Imperative*) is no more a restriction on human action than the objective nature of logic or of the physical world. Rooney, 'Divine Commands', 130.
12 Kant's 'transcendental idealism' does not conflict with 'empirical realism'.

relationship between human freedom and the order in creation envisaged in this approach.

Definition and history

Divine command morality is the name applied to moral systems and theories where the expressed will of the deity is considered the source of obligation.[13] Its modern roots in Christianity lie in the dialectic theology of the 1920s, 1930s and 1940s with Karl Barth and, although the emphasis in his work changed, with Dietrich Bonhoeffer. Barthian ethics epitomises this approach in that it asserts that 'moral imperatives are occasioned by the command of God which governs human existence'.[14] Theological thought can legitimately make reference to claims made by philosophical ethics from this perspective, and positively appropriate representations of human affections and agency, since the doctrine of creation provides the context in which the sharing of insights can be recognised. Along with all Christian theists the proponents of divine command morality are of the conviction that God is the author of all goodness.[15] In addition, although divine command morality does not emphasise the goodness of the created order – in its insistence on the radical otherness of God in his goodness – it cannot discount that goodness entirely. However, the doctrine of creation would not be used as grounds, by the divine command ethicist, for saying that Christian ethics can speak for the whole of humanity. More often, it implies that its message is confined to Christians or perhaps even only to some particular form of church.[16] Although creation theology is the context for any universal ethic, nature is so fallen that

---

13   Cf. Rooney, *'Divine Commands'*, 117.
14   David S. Fergusson, *Community, Liberalism and Christian Ethics* (Cambridge: Cambridge University Press, 1998), 23.
15   Rooney, *'Divine Commands'*, 121.
16   Michael Keeling, *The Mandate of Heaven: The Divine Command and the Natural Law* (Edinburgh: T. & T. Clark, 1995), 2.

human reason cannot be trusted and needs the support of revelation to come to ethical conclusions that are theologically valid.

Thus divine command morality emphasises the place of revelation in moral deliberation. This is not an argument against reason but one in favour of revelation. This constitutive link to scripture does not rule out a nuanced, or historical–critical understanding of scriptural interpretation. It allows that the meaning of scripture is far from self-evident and we often must interpret it. However, it does contest the notion that other positions in theological ethics do 'greater justice to the functions in morality of reason and conscience' or that obedience to God's command is incompatible with the exercise of reason and conscience.[17] It rightly defends the understanding that 'religious' considerations in ethics do not amount to heteronomy. Oliver O'Donovan, for example, defends evangelical ethics from exaggerated dichotomous characterisations. In particular in *Resurrection and Moral Order* he wants to find a route to re-instate the value of the 'created order' so that it is not simply subordinated to redemption in his evangelical ethics.

Scripture as the source of Christian distinctiveness in ethics

Oliver O'Donovan's evangelical ethics can be characterised as a divine command approach but it is also a reconstruction of that position. He does not doubt that for morality to be truly Christian it must accord with scripture. 'Christian ethics must arise from the gospel of Jesus Christ. Otherwise it could not be Christian Ethics'.[18] This may seem unsurprising for Christians and non-Christians alike. Yet O'Donovan wants to say something more specific about Christian ethics in relation to other ethical systems. He wants to make a strong distinction between specifically Christian ethics and values derived from other convictions. So 'Christian believers may have ethical convictions which do not arise from their faith

---

17 Rooney, '*Divine Commands*', 134.
18 Oliver O'Donovan, *Resurrection and Moral Order* (Leicester: Apollos, 1994), 11. Further page numbers are in the text.

in Jesus. There can be ethical Christians without there being Christian ethics' (11). However, he suggests that there is a particular and clearly distinguishable Christian ethic apart from all other ethics, even if some Christians derive their ethics from other sources.

In the preface to *Resurrection and Moral Order*, he expresses his frustration at the 'anti-foundationalism' prevalent in Christian ethics. Apologetic strategies in ethics that work on a simple divine command theory do not do justice, he says, to the moral theological work. Clearly 'proof-texting' is not his objective. He sets out to provide what he calls a systematic rather than apologetic outline to Christian ethics and to trace Christian moral concepts back to their roots in biblical exegesis. O'Donovan does not, however, feel the need to endorse or to subscribe to contemporary traditions of biblical study (11). In the prologue to the second edition of *Resurrection* he locates his position in ethics with regard to some current alternatives, not least the three alternatives explored here – natural law, communitarianism and autonomy.

I will treat Oliver O'Donovan in this analysis as an example of 'divine command' morality in this nuanced sense only. He is not a 'biblicist', nor does he simply subordinate creation to redemption. As we will see, he describes his ethic as a rapprochement of a 'realist' and an 'evangelical' ethic. He does a service to Christian ethics in his exploration of how revelation shapes Christian perspectives in ethics. He suggests that it 'constrains' or obliges Christians to put certain questions to ethics and in that way it relativises any commitment to a particular philosophical ethical system. Nevertheless, even given that Christian ethics is a form of 'theonomy', we cannot simply conclude, on his reckoning, that it is therefore a form of heteronomy. Unintentionally, perhaps, O'Donovan suggests how it is that an autonomy position in ethics can be compatible within a Christian faith context. He sees the need for a 'realist' principle alongside an 'evangelical' principle in ethics. It is then through a focus on the resurrection, which, he says, vindicates the original goodness of creation, that he integrates creation and redemption, autonomy and theonomy.

O'Donovan insists that a 'belief in Christian ethics is a belief that certain ethical and moral judgements belong to the gospel itself' (12). He justifies his ethic – that is, 'realist', 'evangelical' and, more specifically, a 'resurrection' ethic – as follows:

> Purposeful action is determined by what is true about the world into which we act; this can be called the realist principle. That truth is constituted by what God has done for his world and for humankind in Jesus Christ; this is the 'evangelical' principle. The act of God which liberates our action is focused on the resurrection of Jesus from the dead, which restored and fulfilled the intelligible order of creation; this we can call the 'Easter' principle. (ix)

He chooses his emphasis carefully to register that each of these contentions has been negatively challenged, or in some way qualified, in the recent literature of Christian ethics. Although he does not specifically state it, he implies that the realist principle is challenged by communitarianism, the evangelical principle by revised natural law ethics, while the Easter principle is challenged by both. He distinguishes his position from the natural law approach of John Finnis, which, he suggests, makes a good case for morality, but a morality without 'evangelical' content (xi). Finnis's 'reality' principle loses 'purchase' on the evangelical for O'Donovan. He also distinguishes his own ethics from that of Hauerwas who, he claims, isolates the role of witness from other 'Christological moments' in his ethics.

O'Donovan wants to overcome the confrontation between advocates of 'creation ethics' and advocates of 'kingdom ethics'. There is no conflict, for him, between these two:

> A kingdom ethics which was set up in opposition to creation could not possibly be interested in the same eschatological kingdom as that which the New Testament proclaims. ... A creation ethics, on the other hand, which was set up in opposition to the kingdom, could not possibly be evangelical ethics, since it would fail to take note of the good news that God had acted to bring all that he had made to its fulfilment. (15)

He defends the evangelical principle while seeking a realist principle that will take full account of the creation. He defends the autonomy of creation as a gift already given.[19]

Yet, that does not imply for him that the order of creation is self-evident. The creation can only be truly understood, he specifies, from the perspective of the resurrection. He uses a Christological principle to integrate his evangelical and realist approach. He is careful not to isolate the resurrection from other 'Christological moments' of creation, reconciliation and redemption.[20] What can be said in particular of the resurrection is that God stood by his creation and did not allow it to be brought to nothing (xviii). He admits that the text supports and affords us more than one motive for Christian obedience. For O'Donovan it is the task of theology to uncover the relation of things that give the gospels their ethical appeal. For him it is the resurrection that vindicates the creation; it is the new affirmation of life that transforms the gift initially given (14).

In this way O'Donovan's work exemplifies and expands on the divine command approach. The Gospel is more than an ethical textbook but it is nonetheless ethically relevant. It is necessary theologically to make explicit the hermeneutical principles with which we interpret it: as narrative representations of reality; historico-critically; using a Christological principle; assuming a realist principle in epistemology; or as a critique of human justice and politics. O'Donovan follows Barth, but he distinguishes between following him in his epistemology and following him in his ontology. O'Donovan, in contrast to Barth, does not subordinate the doctrine of creation ontologically to Christology. How we know things as humans may be compromised by our fallenness but that does not erode those aspects of creation theology that upheld the original and ongoing goodness of the created order.

---

19   Ibid., 14.
20   O'Donovan privileges the resurrection in contrast to Hauerwas who has the 'tendency to privilege the crucifixion over the other moments of the Christ-event, in keeping with an emphasis on martyrdom and death as the normative expressions of Christian witness'. Ibid., xv.

> In pursuit of an uncompromised theological epistemology Barth allowed himself to repudiate certain aspects of the doctrine of creation ... which ought never have fallen under suspicion ... All this left him with a formal account of the theological basis of ethics which, depending exclusively on the divine command – interpreted in the existentialist way as particular and unpredictable – was far too thin to support the extensive responsibility for moral deliberation. (87)

O'Donovan follows Barth's critique epistemologically; the goodness of created reality is not self-evident to the fallen human person. Nonetheless even in our moral disorder, he says, we do not become swine, like Odysseus's sailors (87). The image of God in us as created may become defaced but it cannot be lost or erased.

A theological deontology as the foundation for ethics

Divine command morality formulates its ethics in terms of a theological deontology. The Word of God and not reason is the final arbiter, and ethical injunctions are framed as duties and obligations, not dependent on outcomes or goals. Divine command provides a non-contingent ethics of duty and obligation as opposed to a teleological one of ends or consequences.[21] Divine command morality is sceptical about human ends or definitions of the good because of its insistence on the effects of fallenness on human nature.[22] As a divine command deontology it stands in opposition to the deontological moral philosophy of Kant – which is based on the autonomy of the subject – and apart from the theological deontology inherent in natural law theory. Rooney argues, for example, that neither Kantian ethics nor natural law theory can do greater justice to the functions of reason and conscience in moral action than divine com-

---

21 However, Bretherton refers to O'Donovan's ethics in this regard as an 'eschatological teleology'. Cf. Luke Bretherton, *Hospitality as Holiness; Christian Witness Amid Moral Diversity* (Aldershot: Ashgate, 2006), 82.
22 This is not interpreted by O'Donovan as radically as in Luther. O'Donovan, *Resurrection and Moral Order*, 64.

mand theory does; and divine command has the advantage of remaining loyal to its specifically Christian remit.[23]

O'Donovan, sees the split between deontology and teleology as somewhat artificial; for him duty and the quest for the flourishing life are not always in conflict. This insight is in keeping with Mieth's theological reconstruction of Kantian autonomy; which also insists on a reintegration of questions of duty and fulfilment. O'Donovan, however, retains a theocentric foundation for his moral deontology whereas Mieth retains Kant's distinction between a philosophical foundation and a theological context for morality. Mieth suggests that it is the Christian context that intensifies, motivates and helps us to discover what is morally relevant. It also integrates insights across disciplines and relativises morality, as we shall see.

For O'Donovan the 'modern obsession' with the deontological/teleological split in moral philosophy sees all of our duty or our obligation as a burden laid upon us; a burden that cuts across our natural aspirations and purposes. In contrast he suggests that these are two languages that draw our attention to different but complementary aspects of moral claims:

> The deontic language emphasises the critical relation of moral authority to natural authority and of divine authority to all created authority. We say 'ought' when we need to stress the contradiction between this overriding claim and what we would have otherwise thought or felt. Although it is mistaken to claim that only the conception of obligation qualifies thought as truly moral, this mistake conceals a truth: for it is certainly the case that we think morally only as we think critically, and the force of the 'ought' is to stimulate us to do that. On the other hand, the teleological language draws attention to the rationality of moral and divine authority. It drives us to express, as best we can, the meaning of that authority within the ordered universe, even when that expression is the expression of hope for a resolution that we cannot yet comprehend. (139)

Deontological language alerts us to the duty to which we are called, moral and theological, against what we might have otherwise assumed simply to be obvious and clear. Teleological language, in contrast, stresses

23 Rooney, *'Divine Commands'*, 134.

the connection, not the disjuncture, between what we experience and what we validate and re-imagine, in realising our ethical obligations. O'Donovan, and Mieth, see the possible compatibility between faith and reason, duty and happiness; they are not simply always in conflict, as I hope to show.

However, for O'Donovan, the call to our duty comes from knowing God as God has revealed God's self in scripture, rather than through an analysis of practical reason. The 'meaning of that authority' and 'the expression of a hope for a resolution' refer to Christian truth claims and Christian hope; the claim that divine authority can bring about a hoped-for resolution of the 'antinomy of practical reason', where fulfilment and duty coincide. Kant, as O'Donovan reminds us, also re-instates a notion of the 'highest good' in his 'postulates of practical reason': God, happiness, immortality.

> ... the notion of the 'highest good' is reinstated in Kant's ethics as the object, although not the motive, of the pure practical reason, and we learn that the Christian eschatological hope of the kingdom of God, complete with the idea of happiness, is, together with the immortality of the soul, a necessary 'postulate' of practical reason. (139)

For Kant the grounds of obligation are to be found in the autonomous subject not God. God is the answer to the subject's quest for meaning and when our best intentions founder in a hostile world. God, freedom and immortality are postulated by practical reason at the 'limits of reason' and are sources for moral motivation, not the foundation for our obligation. O'Donovan, not surprisingly, protests that Kant is trading in 'unconfessed theological contraband' in importing theological ideas of Christian eschatological hope and the idea of happiness into his moral philosophy (263). Mieth, however, is not convinced by this argument and argues instead that Kant is rightly following his philosophical analysis to its conclusion; at the limits of reason, we come to the antinomy of practical reason and to the question of meaning and of hope. Philosophically it is not a question of aligning ourselves with the 'order of the universe', it is a question of a free response to our moral duty from an analysis of practical reason. Theologically it is not a question of aligning ourselves with the authority of the divine will, but a question of a free response to God's offer of divine fellowship. It is part of the meaning of Christian hope that our fulfilment and our duty are reconcilable, even if that is not evident in our own experience. Mieth might well concur with O'Donovan that in

> ... a moment of grace we may be given the perception that our duty and our fulfilment are one and the same, and we may speak of that unity in hope and faith; but we cannot ask that we should never be challenged to further thought and conscientious struggle by an awareness of the divergence of inclination and duty. (139)

Mieth however would differ from O'Donovan in that he would offer a philosophical foundation for morality in his reconstruction of Kant, even if that foundations is specified and relativised in a Christian context.

Divine command morality provides a deontological foundation for ethical behaviour, as does Kantian ethics. O'Donovan recognises a degree of artificiality in the split between deontological and teleological aspects of moral claims in ethics. O'Donovan's ethic is, however, wholly theological on the question of the foundation of moral obligation. This theological deontology is in direct contrast with the philosophical deontology of Mieth's Kantian ethics. O'Donovan's reconstruction has implications for the question of how ethical claims from Christian voices are formulated and expressed in the public realm.

A specifically Christian discourse

It seems that divine command morality's use of scripture, as the point of departure for ethical argumentation, leaves it open to the charge of exclusivity and particularity, indeed to being esoteric and inaccessible outside of Christian contexts. So interpreted, it could be said to have a limited commitment to public dialogue. However, there is a danger that such a characterisation of divine command morality too easily becomes a caricature. Divine command morality defends 'taboos' that have grown out of historical experience and these cannot be simply discounted in the public realm just because they originate from a theological source. It is not a contradiction that a caveat in ethics can be both theological *and* publicly relevant at the same time. A 'taboo' may be revisable in light of new evidence, or it might be upheld because it reflects the reality of lived human experience, albeit captured at times in non-conceptual and narrative forms. Such taboos are ethically justified particularly where they are understood as cautionary fences.[24] Ethics cannot rest there of course. It is an ongoing task for Christian theology to mediate the potential of its own categories in ethics and give an account of its sources and value frameworks.[25]

However, it is not theological conviction that is problematic for divine command morality in the public realm, at least not from the perspective of an autonomy perspective in Christian ethics. Rather, the problem lies in its conviction of the fallenness of the human creature and the inadequacies of taking human reason as a starting point in ethical discourse. This conviction lies behind the way it envisages itself as in conflict with a contemporary public morality with its over confident appeals to reason.[26] It takes a 'prophetic' stance, against the darkness of cultural moral relativism, yet in doing so it implicitly, even if inadvertently at times, characterises all non-Christian ethics as 'moral relativism' and all non-Christian rationality as instrumental.

In developing a theology of the public realm, divine command morality, so understood, could take several paths. It could seek the reinstatement of some type of Christendom, or advocate a separatist agenda that disengages with contemporary culture. It could alternatively opt for a radical orthodoxy approach; retreating from the world of politics, at

---

24   Dietmar Mieth, 'Bioethics, Biopolitics, Theology', in *Designing Life? Genetics, Procreation and Ethics*, ed. Maureen Junker-Kenny (Aldershot: Ashgate, 1999), 21.
25   This is true even in the face of what Habermas calls the danger of a 'political overgeneralisation of the secularist world view'. Cf. Maureen Junker-Kenny, 'Valuing the Priceless: Christian Convictions in Public Debate as a Critical Resource and as "Delaying Veto" (J. Habermas)', *Studies in Christian Ethics* 18, no. 1 (2005), 52–54.
26   Cf. H. Richard Niebuhr for a classic presentation of the enduring tensions and relations between faith and culture. H. Richard Niebuhr, *Christ and Culture* (New York: Harper and Row, 1951).

once construed as hostile, into the development of a Christian ecclesiology. This is the temptation for Christian communitarians. These are paths, however, that O'Donovan implies he would specifically seek to avoid.

The symbolic end of Christendom came, for O'Donovan, with the First Amendment to the Constitution of the United States of America.[27] The separation of church and state suggested that the state should not offer deliberate assistance to any of the churches in their mission. The state was 'freed from all responsibility to recognise God's self-disclosure in history' (245). However the outcome of this is that something that was conceived of as a liberation from the straits of religious government became a tool of antireligious sentiment. Instead of a continuation of salvation-history we have a kind of 'unfolding providence' and thereby a Christian secular government that is no longer intelligible (246).

The reinstatement of Christendom is not something commonly appealed to in modern pluralist contexts. Such an ambition has been considered by O'Donovan in his *Desire of the Nations*.[28] His more modest objective as regards 'political theology' is not so much to reinstate a kind of Christendom but to claim a stake in the achievements of culture that owe a debt to Christianity, and to have that debt acknowledged. For O'Donovan the 'false self-consciousness of the would-be secular society lies in its determination to conceal the religious judgements it has made' (247). The alternative to Christendom or separatism, for O'Donovan, appears to be the path that continues to trace how liberal society has been influenced, and is still influenced, by its Christian roots. In 'tracing the characteristic features of liberal society, then, we need to show how it has been affected by the narrative structure of the church, which is itself a recapitulation of the narrative structure of the Christ-event' (250).

Nonetheless, he continues to stress the fallenness of human reason and the necessity of the divine command for a political theology. He goes on to say that the idea that social order is based on a single principle, such

---

27 Oliver O'Donovan, *The Desire of the Nations* (Cambridge: Cambridge University Press, 1996), 244.
28 Cf. Ibid.

as liberty, is ideological, and a political theology cannot be ideological. For this reason a theology of the public realm must give some content to what it understands by freedom by appealing to 'divine command'. A 'political theology which defends freedom without filling it out with the content of the divine command and the divine redemption of society in Christ is ideologically liberal' (250). So for O'Donovan only divine command can free human reason and freedom from ideology.

Yet there are interpretations of secular reason, freedom and society that warrant a more positive theological evaluation and which are not easily reduced to being 'ideologically liberal'. A theology of the public realm could be predicated on a more positive theological interpretation of the autonomy of the human person and of creation; namely, the creation of finite human subjects as addressees of God's self-revelation and intention for the autonomy of creatures. This will be explored further when I come to discuss the autonomy approach.

Non-human creation from a divine command perspective

O'Donovan has paved the way for a consideration of the value of the creation, and by implication non-human nature, by declaring his ethic a 'realist' ethic. He is not endorsing 'empirical realism' here but rather endorsing the value of creation, theologically understood. Creation, it is suggested, has too often been 'subordinated' to redemption in divine command morality. His theme is to rehabilitate creation, alongside redemption, in his ethics. However, the 'order of creation' is never self-evident; we remain in need of a 'kingdom ethic' so we are not otherwise confused. As regards the non-human creation, O'Donovan specifically says that we do have biblical warrant (Rm. 8:19–24) for holding to the hope of redemption for all 'creation', even if it not clear what that may be (54).

The strength in O'Donovan's approach, for Christian ethics, stems from and is consistent with the sharp divide he draws between begetting and making, between God's creativity and mere human making.

> If there is no category in thought for an action which is not artificial, then there is no restraint in action which can preserve phenomena which are not artificial. This imperils not only, or even primarily the 'environment' (as we patronisingly describe the work of things that are not human); it imperils what it is to be human, for it deprives human existence itself of certain spontaneities of being and doing, spontaneities which depend upon the reality of a world which we have not made or imagined, but which simply confronts us to evoke our love, fear and worship.[29]

It is in this critique of a 'technological mentality' that O'Donovan's ethics contributes to the debate about the non-human creation. He privileges the 'givenness' of creation while at the same time he critiques the valorisation of human 'making'.

Nevertheless that does not mean that we cannot make use of creation, even if we need Christian love to avoid 'playing God'.[30] In keeping with the Augustinian tradition, he suggests that the use of reproductive technologies, for example, in farmed animals may be authorised for the good of humanity. Non-human nature is disposable to the ends of society and harm may be done to it for the human good.[31] In reference to outer nature he claims that 'although cattle, like human beings, live individuated lives which are extended through time, there is no particular significance which resides in the individual life course of each'.[32] It is the question of what it is to be human that is his most immediate major concern in light of dilemmas raised by reproductive technologies.

Clearly authors use the category 'nature' in radically different ways. O'Donovan is using 'nature' theologically as a specification of what it is to be a human creature in the first instance, and as an extra-human phenomenon in the second. O'Donovan's concern, at least in part, is to preserve phenomena not because they are human, or for human use, but because they are natural, spontaneous. Christian scripture and theology tell us, he says, that the creation is not intended for annihilation or repudiation.

---

29   Oliver O'Donovan, *Begotten or Made?* (Oxford: Clarendon Press, 1984), 3.
30   Ibid., 83.
31   Ibid., 49.
32   Ibid., 51.

In the first place it rules out that threatened end of all things which is implied by the Fall; corruption and disintegration. It rules out also the gnostic possibility that creation is to be repudiated or overcome in the name of some higher good. It rules out the possibility that the temporal extension of creation forms a random and meaningless flux.[33]

The question remains as to how it is that humanity participates in creation? For O'Donovan humanity participates in the created order through morality. 'Morality is man's participation in the created order' (76). O'Donovan also recognises that humanity participates in the created reality through knowledge. 'Knowledge is the characteristic *human* way of participating in the cosmos' (81). However, he frames Christian morality in terms of the human person's freedom to 'conform' to a prior order. 'Christian morality is his [Man's] glad response to the deed of God which has restored, proved and fulfilled that order, making man free to conform to it' (76). And he stresses that human knowledge is mistaken, confused and deceived. A technological mentality has led us, he says, to a 'confusion' in relation to food and the 'environment'; particularly when we begin to assume that we have devised the world for our own purposes. In relation to food, he says, that the human person is so confused that he 'continues to eat vegetables; but he no longer knows that he does so because they *are* food, and comes to imagine that he has *devised a use* for them as food' (52). It is this confusion, moral and cognitive, that O'Donovan's divine command positions holds out against.

Summary

Superficially, divine command ethics is the position from which arguments against the dangers of 'playing God' are made and in which a sharp dichotomy is drawn between Christian and non-Christian culture. For O'Donovan this is a parody of the divine command position. The strength

---

33 O'Donovan, *Resurrection and Moral Order*, 55. Further page numbers are in the text.

of his position lies in the narrative resources that it brings into play that express concerns not always obvious in conceptual discourse. O'Donovan wants to defend a theological imagination that does justice to a theological and scriptural view of reality; the view that the reality we encounter is the creation of a good God. For O'Donovan, it is the life-giving spontaneity of the natural, in contrast to the dull predictability of the artificial, that evokes our reverence and awe.

Divine command morality insists that liberal society owes a debt to Christian theology and ethics even if it fails to recognise that debt. This does not mean to suggest, however, that Christian ethics speaks for the whole of humanity. Yet, even if O'Donovan's ethic does not start from an analysis of universal human reason, he does not confine the relevance of his ethics to some notion of church. His stated aim is to keep both a 'realist' and an 'evangelical' principle at play in his ethics. In this way his ethical position can have recourse to non-theological disciplines even if he does not always make these explicit.

Divine command morality provides a theological deontological foundation for ethics. Our duty is predicated on the expressed will of God and does not spring from human reason or human inclination. However, O'Donovan insists that human morality and knowledge are the means through which humanity participates in the creation, even if we must always be reminded that they are finite and fallen. Theological ethics, however, cannot be equated with heteronomy. Theonomy and autonomy are compatible. The question of how they are compatible will be taken up in the discussion on autonomy in Christian ethics. In contrast to O'Donovan, the autonomy position grounds moral obligation philosophically and not theologically, in its Christian reconstruction.

Finally, although O'Donovan defends the 'givenness' of creation against a 'technological mentality' it is not obvious how his position would develop as regards non-human nature. He is not averse to the 'use' of natural resources for the human good. Humanity 'participates' in the creation through morality and knowledge and even if the human person is morally weak and our knowledge finite and fallen, humanity still has dominion (81). This emphasis on human moral responsibility before God echoes the emphasis that the autonomy approach places on human moral obligation. In chapter four, *Nature in Theological Perspective*, I will examine how O'Donovan's ethical position has been put to work by Michael Northcott in relation to environmental questions and agricultural production in particular. As regards non-human nature, O'Donovan's strength is that he defends both the goodness and givenness of the world that we inhabit before he defends human creativity. It is only in the context of understanding creation as a gift, a gift that is not of human making, that we can then talk of the ethics of human activity, knowledge and morality in the world.

In the next section I examine communitarian perspectives in theological ethics. These are also critical of human reason but place their emphasis not so much on the Word of God in Scripture but on the role of Christian communities in moral formation.

Christian communitarian or ecclesial ethics

Communitarianism is the term applied to a movement in social philosophy in the 1980s that suggests that our morality should be approached from the perspective of the moral community to which people belong. Its appearance may be traced to a series of publications that launched an attack on modern liberal individualism and atomism.[34] *After Virtue* by Alasdair MacIntyre is taken as a decisive text. Communitarianism has also been credited with the revival of Aristotelian virtue ethics, and ecclesial communitarianism with a revival of virtue ethics interpreted through the theological virtues of faith, hope and charity. This revival is principally envisaged as an alternative to, rather than as a complement for, deontological approaches. However a dichotomy between deontological and teleological aspects is not necessary to a virtue approach. Indeed virtue ethics

---

34 David Fergusson, 'Communitarianism and Liberalism: Towards a Convergence', *Studies in Christian Ethics* 10 (1997).

makes a significant contribution to the rapprochement of the two. At least from an autonomy perspective it is clear that deontological ethics needs to be supported at the level of values.³⁵ Likewise, as Junker-Kenny points out, in her discussion of Joas's *Genesis of Values*, the 'particularity' of traditions which enable experiences of 'self-transcendence and self-transformation' are not always at odds with universal principles. In the work of Charles Taylor we find a communitarian approach that is not unfriendly to the achievements of modernity but nonetheless wants to build a less abstract anthropology than can be found in atomistic individualism.

Communitarians read individualism, since it leaves out community, only as a truncated and inadequate anthropology, although communitarianism itself has no shared anthropology or model of society. Indeed it appears to be united predominantly by its critique of liberalism.³⁶ As an ethic that stands in opposition to any hint of the Enlightenment ideal of universal morality it has had to face the criticism that it is antirational and sectarian. Its emphasis on vital traditions, however, can be seen as a helpful correction of the individualist denial of the need for cultural traditions that foster values. A retrieval of the role of community in moral formation where it complements rather than conflicts with autonomy, has much to offer anthropology and ethics. It recognises that being part of a community is constitutive of human reality and that community and individual are not always or necessarily in conflict. It suggests for example that there are intermediate institutions between the state and the individual beyond and above economic and market mechanisms. In its ecclesial form it suggests that a Christian way of life is grounded in an ecclesial community and by virtue of that is not simply a private affair. The church, as an intermediate institution, represents and mediates for communities, although this mediation is less effective when viewed from a more polarised ecclesial communitarianism, such as is found in the radical orthodoxy movement. Radical orthodoxy notwithstanding, the Christian churches have traditionally defended a concept of the person

---

35 Junker-Kenny, 'Valuing the Priceless', 50.
36 K. Grasso, G. Bradley, and R. Hunt, eds, *Catholicism, Liberalism and Communitarianism* (Lanhan, MD: Rowman and Littlefield, 1995), 4.

in society. They are well placed, in that sense, to contribute to the institutional framing of policy, in the public realm.

Communitarianism has many positive implications in areas of Christian ethics where an emphasis on community and on fostering moral character is reintegrated – not set up in conflict with – an anthropology of 'situated freedom'.[37] The emphasis on church as a community of interpreters also adds a hermeneutical 'extra'. It sees the interpretation of scripture as a question of the history of reception and interpretation of biblical texts in local church communities world-wide.

In a similar way communitarian themes have the potential to contribute to the development of an 'environmental virtue ethics' in that they foster a community perspective on and appreciation of the 'global commons'. This appreciation extends beyond the issue of the individual benefits of, or interests in, our common natural heritage to questions of intersubjective generosity and institutional justice. A revival in an 'ethics of tradition', an ethics of the good life and in 'virtue ethics' – a revival that can be traced in German-speaking theological ethics to the work of Dietmar Mieth and in the English-speaking world to the work of Alasdair MacIntyre – have much to contribute to environmental ethics. From an autonomy position, of course, any new-found beneficence towards nature needs to be complemented by a deontological framework so that concepts of the good, on which the virtues depend, do not themselves become tools for instrumental reason.

Definition and history

The strength of communitarianism, as a social philosophy, is not as much in its critique of liberalism as its critique of liberal 'atomism'. Atomistic individualism can only deliver an abstract, partial, asocial and decontextualised view of the human person. Atomism 'posits the self as given, prior

---

37 Charles Taylor, *Sources of the Self: The Making of Modern Identity* (Cambridge: University Press, 1989), 515. Further page numbers are in the text.

to any social order – ahistorical, unsituated, a bearer of abstract rights'.[38] It is perhaps no coincidence that this critique is loudly articulated in North American moral philosophy and theology where the cultural history of that society leads it to identify atomistic individualism as *the* danger to social coherence. In Europe, cultural history resists a different danger, that of totalitarianism. Lisa Sowle Cahill, an American moral theologian and advocate for the Catholic social doctrine of the common good, in dialogue with the German theologian Dietmar Mieth, remarked on:

> ... the importance of presenting a theory of the common good in a manner attractive to those whose cultural history posed totalitarianism as a greater social danger that the individualism that I as a North American theologian had been trained to battle.[39]

As a critique of 'atomistic' individualism, communitarianism wants to build a less abstract picture of the human person. It registers the philosophical and theological critiques of unsituated human agency, of the disengaged subject since Descartes.[40] For Charles Taylor the 'punctual' self is part of an 'invalid' anthropology and view of agency. It is an anthropology that is not even necessary to the pursuit of self-responsible reason and freedom, although it is often painted as being such (514). The point, for Taylor, is to develop 'anthropologies of situated freedom'. In doing this he does not seek to replace the hegemony of atomism with that of pluralism – one anthropology with many anthropologies. Or indeed to replace it with relativism, in which all anthropologies are equally valid. The more significant point he makes is that not everything we do is at the level of duty. We have duties and obligations, which are universalisable,

---

38 Jean Bethke Elshtain, 'Catholic Social Thought, the City and Liberal America', in *Catholicism, Liberalism and Communitarianism*, ed. K. Grasso, G. Bradley, and R. Hunt (Lanham, MD: Rowman and Littlefield, 1995), 100.

39 Lisa Sowle Cahill, 'Genetics, Individualism, and the Common Good', in *Interdisziplinäre Ethik; Grundlagen, Methoden, Bereiche; Festgabe Für Dietmar Mieth Zum Sechzigsten Geburtstag*, ed. A. Holderreg and Jean-Pierre Wils (Freiberg: Herder, 2001), 378.

40 Taylor, *Sources of the Self: The Making of Modern Identity*, 514.

but we also have visions of the good life and of beneficence, of a plurality of goods. It is in the context of our visions of the good life that we can acknowledge and realise our freedoms and duties.

Given that, Taylor is aware that it is no simple matter to adjudicate between visions of the good. We cannot ignore that the plurality in visions of the good life come with concomitant conflicts and tensions (518). The problem, for Taylor, is not that conflict arises but that it is seen as inevitable. It is taken as a self-evident truth that any presentation of one vision of the good life requires us to mask or delegitimise another, whether we argue from a secular or a theistic position. The proposed solution to the problem of potential conflict is to defend 'pared-down' anthropologies and visions of the good life that might be acceptable to all because they favour none. However, even a 'stripped-down secular outlook' can also cost too much, anthropologically speaking, just as a the 'great spiritual visions of human history' have sometimes laid crushing burdens on humankind (519–520). Instrumentalist anthropologies that discount 'visions of the good life' or 'fulfilment' or 'ties with nature' ignore the costs because they do not or cannot recognise them (518). This is the strength of a communitarian critique. That, of course, is not to say that communities have no shortcomings, Taylor does not fall into that trap. He is aware that living in community has its own limitations. Triumphalist communitarianism, like atomism, draws legitimate moral and ethical critique.

Church as the source of Christian distinctiveness in ethics

Ecclesial communitarianism focuses not so much on the Word of God in scripture, but on the Word of God in the community. It is the church, as the institutional community of interpreters of scripture, that distinguishes Christian ethics from 'liberal' or other 'secular' ethical positions. Ecclesial communitarianism can draw on a variety of sources.[41] With a

---

41 G. Paul, 'Communitarianism', in *International Encyclopedia of Ethics*, ed. John K. Roth (London: FD, 1995), 172.

focus on community it is free to draw on the resources of divine command positions without compromising its central tenet; community as authority and judge of sources. It can also draw on post-modern, post-liberal approaches to biblical exegesis and theological doctrine, like those of Hans Frei and George Lindbeck. Frei advocates a narrative interpretation of scripture while Lindbeck develops a cultural-linguistic theory of doctrine; understanding Christian doctrine is akin to learning the grammar of your native language. This rule theory envisages doctrine as a language which is best learned and understood in relation to the community in which it is spoken, as opposed to propositions rigidly adhered to or propositions that are interpreted merely as particular expressions of a universal religious experience. The value of the communitarian retrieval of narrative theology and virtue for an ethics of the good life, that takes us beyond questions of duty, cannot be underestimated.

Communitarianism, however, risks being construed as non-cognitivist, or a 'regulative' rather than 'realist' account of faith.[42] It can imply that there is no 'objective' truth in morals, only constructed truth. The ecclesial community, in that view, constructs and judges truth. 'Truth in morals is thus constituted by reference to the beliefs and practices of whichever community and traditions one owes allegiance to and there is no transcommunitarian fact at stake into which one can reasonably inquire'.[43] In this way a communitarian ethic may satisfy the demands of internal consistency but can fail to address the question of the general consciousness of truth in morals.

Drawing a sharp dichotomy between secular liberal society and the Christian community, as Hauerwas does, presents dangers for an ecclesial ethics. The temptation is to dismiss secular reason because it is assumed to be part of fallen reason. Over-stressing the role of the community can serve to undermine all of the hard-won achievements of modern theology and unnecessarily deepen and solidify the differences between secular and Christian discourse. The worst excesses of secularism are contrasted with

---

42 Fergusson, *Community, Liberalism and Christian Ethics*, 53.
43 Ibid., 7.

the finest achievements of Christianity and neither side acknowledges their debt to one another in mutual critique.

Communitarianism also risks devaluing secular anthropologies unnecessarily. Stanley Hauerwas, for example, claims that the human person is only truly human when Christian. He over-dramatises the crisis of liberalism. Symptomatic of this is his exaggeration of the differences between Christian and non-Christian morality. On the question of marriage and parenting, for example, his accentuation of the difference alienates not only secular culture but also other Christian ethical positions. As Fergusson suggests, this 'places him at odds with both the Catholic and Protestant traditions which have seen these institutions as wider in scope than the church and have detected good practice in non-Christians as well as Christians'.[44]

Christian morality and Kantian liberalism, as Hauerwas understands it, are simply incompatible. Indeed Christian faith and autonomy are not only incompatible but are at odds with each other. The 'kind of life Christians describe as faithful is substantively at odds with any account of morality that makes autonomy the necessary condition and/or goal of moral behaviour'.[45] In this way his particular type of communitarianism renders liberal allies as foes and hinders the process of making common cause. Yet it is possible to provide a more positive theological warrant for autonomy, based on the prior unconditional gift of human freedom and the offer of divine fellowship with God.[46] Kantian ethics shares more with Christian ethics than Hauerwas would allow for – in terms of protecting the dignity of the human person against commodification for example – and this is the case even outside of a specifically theological reading.

Lastly, the Jesuit dogmatic theologian Georges De Schrijver questions the bona fides of attempts made by some communitarians to 'save

---

44 Ibid., 73ff.
45 Jean Porter, *The Recovery of Virtue: The Relevance of Aquinas in Christian Ethics* (London: SPCK, 1994), 23.
46 A comparison of MacIntyre's anti-modern approach, which Hauerwas follows, and Charles Taylor's more friendly approach is beyond the scope of the present enquiry into communitarianism.

orthodox Christianity with the help of postmodern narrativity'.[47] He suggests, for example, that John Milbank, one of the authors of the Radical Orthodoxy movement, uses 'narrative' in a way that goes beyond what theologians usually mean by narrative theology and by the church as a community of interpreters. It is not the focus of this study to evaluate the problems and potential contribution of narrative theology to Christian ethics. The communitarian authors presented here would, no doubt, stress different aspects of narrativity in their ethics; the openness that narrativity brings, the 'fusion of horizons', or indeed harmony and peacefulness. In this study I want, rather, to take up a critique of the concept of narrativity at work in radical orthodoxy. This is because Milbank's view of narrativity represents a danger for any approach (including that of autonomy in Christian ethics), that sets out to recover an ethics of tradition and of the good life as lived out in the Christian community.

Narrative theology, for De Schrijver, is usually concerned with a particular task, specifically with the critical re-appropriation of the tradition. This 'reappropriation of tradition is meant to bring about a fusion of horizons ... between tradition and modernity, whereby the recounted narrations of the past are expected to add a specific content or important nuance to a discussion', according to De Schrijver (44). Milbank's 'narrativity', he suggests, means something else. It is closer to an understanding of narrative as 'fiction' as 'a product of creative imagination people need to give meaning to their lives in terms of aesthetic consolation.' (44) In the postmodern 'agonistic' clamour each perspective must compete in an open contest with other voices. For Milbank the 'struggle' is what marks the 'pagan milieu' and it is in this context, for him, that the contrasting irenic nature of the Christian vision emerges as superior to all others.[48] For De Schrijver Milbank's understanding of narrativity exaggerates the differences between modern secular and 'postmodern' Christian positions.

---

47 Georges De Schrijver, *Recent Theological Debates in Europe: Their Impact on Interreligious Dialogue* (Bangalore: Dharmaram Publications, 2004), 44.
48 Cf. John Milbank, 'The End of Dialogue', in *Christian Uniqueness Reconsidered; the Myth of a Pluralist Theology of Religions*, ed. Gavin D'Costa (New York: Orbis, 1997).

Certainly from an autonomy perspective it is possible to evaluate modernity more positively, not as a Christian heresy but as the 'unfolding' of the Christian tradition (37). Narrative theology and the church as a community of interpreters serves this unfolding Christian tradition. Ecclesial communitarianism does not need to ontologise the secular as violent in order to strive for a church of peace. Indeed De Schrijver questions whether Milbank's myth of Christian peace and harmony, snatched from the jaws of pagan agonism, adequately captures the highest good for Christian faith.

For De Schrijver the project of modern theology is at stake in Milbank's critiques. There are, he says, basic patterns in modern theology that he would want to see continued and transmitted, not least 'the view that the human beings are the agents of their own histories, and that this ... autonomy and responsibility is part of their being created in the image of God; ... that religiosity is not just a matter of "Churchy-ness", but rather demands that the deep concerns for which religion stands be extended to the whole of our "being-in-the-world"'.[49]

A teleological ethic

Communitarianism is credited with a revival of Aristotelian virtue ethics, principally as an alternative rather than as a complement to deontological or consequentialist approaches in philosophy. Virtue ethics is built on a teleology or a concept of the good towards which the human person is oriented. To re-appropriate Aristotelian virtue ethics, MacIntyre sought to establish a teleology for a return to communitarianism. As Lisska observes, MacIntyre substituted an attenuated or social teleology for a natural teleology.[50] In this way he did not have to insist on a necessary link between his rehabilitation of virtue and the classical metaphysics of

---

49 De Schrijver, *Recent Theological Debates in Europe*, 122.
50 Cf. Anthony Lisska, *Aquinas's Theory of Natural Law: An Analytical Reconstruction* (Oxford: Clarendon, 1996), 190.

Aristotle (and of Aquinas subsequently) in which virtue ethics grew up. This allowed him to co-opt virtue ethics for communitarianism.

However, MacIntyre became aware of the necessary connection between virtue ethics and the metaphysics in which it is embedded, between the moral claims of natural law theory and the metaphysical propositions that serve as its foundation. He eschewed the inadequate, liberal deontology he saw in Kantian ethics in order to find a communitarian virtue ethic, only to discover the need for a teleology not unlike that found in classical natural law theory. As Lisska already observed, such a teleology functions deontologically because it is based on essential human (rational) nature rather than consequentially on the outcome of moral actions (190). In contrast, in its ecclesial form communitarianism does not leave out the 'metaphysical foundation', which might act deontologically. Rather it collapses the question of the right and the good into a discussion of the virtues making the task of defining criteria in Christian ethics more difficult.

In one sense, undoubtedly, it is uncontroversial that ecclesial ethics declares the church community as the starting point for Christian morality. However, ecclesial ethics cannot rest there, it needs to spell out its view of church so that its ethical criteria can be clarified. The drive to reappropriate Augustine's 'city of God' for an idealised view of the church must raise questions for any theological ethics. Without a complementary deontology, teleologies suffer from what Dietmar Mieth calls the 'motivational fallacy'; confusing a good motive for a good argument. Junker-Kenny says:

> I detect the equivalent of 'wild hopes' in Christian ethics in an equally undisciplined thrust towards defining and achieving the perfect Christian, the ideal church, an absolved conscience, the end of all conflict. From here, it is only a short step to insist that only a spotless church, a sinless disciple, are entitled to voice specific criticisms about secular decisions. Besides being theologically questionable as a flight into a never-realised eschatology, it also constitutes a short-circuit in ethics. Moral authority based on integrity makes critique more credible, of course, but

it is not a condition of the validity of an argument that the motivations of the discourse participants are above all doubt and blameless.[51]

Ecclesial ethics, like other positions in ethics, is not immune, she suggests, from the need to give an account of its sources and value frameworks.

Having said that, communitarianism nonetheless develops the teleological aspects of ethics as a reaction against the overly abstract views of the decontextualised human person characteristic of deontological approaches. The emphasis on community has several advantages, not least that it pays attention to the genesis and renewal of values and it emphasises the social and political aspects of human fulfilment.[52] In its Christian reconstruction the role of the ecclesial community in moral formation is retrieved. It takes Christian ethics beyond questions of duties and obligations, which must be met, to a different level of appreciation for what has been given and the goals and tasks ahead. Striving for the 'good life' well lived, where it may be deontologically neutral, remains teleologically compelling.

However, communitarianism is in danger of over-emphasising the relational over the reflexive aspect of the human person. Without the 'pole of reflexivity' – the concomitant insistence on the dignity of each individual – communitarian one-sidedness promotes social conformity over individual conscience. Atomism is a problem in that it devalues one's ethos but it is important to also see the opposite problem, that conformity to the community's ethos becomes a condition for belonging and akin to an obligation, instead of being freely chosen, as Junker-Kenny concludes. Autonomy is not the antithesis of community but is a condition for the conscious and critical appropriation of one's own tradition.[53]

---

51 Junker-Kenny, 'Valuing the Priceless', 45.
52 Maureen Junker-Kenny, 'Virtues and the God Who Makes Everything New', in *A Wandering Galilean: Essays in Honour of Sean Freyne* (Leiden: Brill, 2009), 6.
53 Junker-Kenny also points out, following Schnädelbach, that communitarianism undervalues the concept of legality, the 'most important non-Aristotelian element in our modern ethos'. Legality is also a condition for personal freedom, it allows a living together that does not require a shared understanding of the good. The

Communitarianism, in prioritising the good before the right, makes the whole of morality teleological, undermining the potential for a mutually supporting deontological framework. Its ecclesial form rightly highlights the social and political aspects of belonging to a community of faith. However, this needs to be tempered by a corresponding defence of the self-reflexive as well as the intersubjective aspects of moral formation in community.

A shared ecclesial ethical discourse

Communitarianism carries with it an implied rejection of a universal moral system based on a shared understanding of human reason or the human good. In its radical ecclesial form it characterises secular morality as agonistic and incompatible with Christian social theory. In this way it alienates potential allies in the public realm unnecessarily and ironically undermines the concept of the common good. However, it is in its contribution to less abstract notions of the human person, that recognise that 'sources of the self' are intersubjective as well as self-reflexive, that communitarian themes are a welcome development in teleological approaches in ethics and a complement for a deontological ethics. Communitarianism can remind liberalism of both the change and continuity in values in society and of the need to engage with the institutions that support those values, of which the church is one.

Although Alasdair MacIntyre allows that the nation state promotes certain goods, he does not allow that it can promote the common good. The state, for MacIntyre, rather than being a keeper of the common good, resembles a large corporation that provides necessary but limited services to the individual. Going to war to protect the nation state, he writes, in a (in)famous quote is 'like being asked to die for the telephone company'.[54]

---

losses in social bondedness implicit in legal regulation are, she says, secured at a different level. Ibid., 6–7.

54   Alasdair MacIntyre, 'A Partial Response to My Critics', in *After Macintyre; Critical Perspectives on the Work of Alasdair Macintyre*, ed. J. Horton and S. Mendus (Notre Dame: Notre Dame University Press, 1994), 303.

Clearly, for MacIntyre the common good, which he values, is more than mutual forbearance and tolerance; which is all that the state can rise to. This leads him to encourage the formation of smaller voluntary associations, such as the churches, to pursue their own good. Ironically he thereby immediately neutralises any commitment to the common good.[55]

Ecclesial ethics, in its radical form, follows MacIntyre's lead but deepens its critique of the liberal state and elevates the institution of church above all others. The radical orthodox (RO) movement in Britain, which can be understood as a strong form of communitarianism, critiques political liberalism as 'ontologically violent and ethically nihilistic'. It argues that only 'participatory theology' can overcome the violence of social atomism.[56] Radical orthodoxy views liberalism not as humanistic forbearance as much as a pacifier for the will-to-power.

Christopher Insole, a theological critic of radical orthodoxy, views this stress on the violence in human society as an excessively 'constructivist' view of truth. It is self-fulfilling, in that the problem – ontological violence – suggests its own remedy – an Augustinian world view based on peace and harmony. 'Radical Orthodoxy not only seems to share this outlook of residing entirely within a nihilistic universe, but exacerbates this condition by asserting and constructing Augustinian worldviews.'[57] The yearning for a more unitary society and the 'strong doctrine of participation' that RO advocates springs directly from the perception that liberal culture is characterised by social 'atomism'. However, such strong doctrines of participation leave little room for conscientious dissent or for the universal message of salvation promised in Christ, as Junker-Kenny points out:

---

55   Fergusson, 'Communitarianism and Liberalism: Towards a Convergence', 45.
56   Christopher Insole understands radical orthodoxy to be a strong form of communitarianism; namely that the individual is constituted largely by their community and that ethics, politics and morality should further the natural priority of the community over the individual. Cf. Christopher Insole, 'Against Radical Orthodoxy: The Dangers of Overcoming Political Liberalism', *Modern Theology* 20, no. 2 (2004).
57   Ibid., 223.

The unity of faith gets stressed to an extent that faithful, practical and theoretical conscientious dissent from what has been defined as orthodoxy and orthopraxy is hardly conceivable. And the possibilities of mutual learning and critique, instead of constant correction of secular society, as well as the chance for the world to repent are no longer held open.[58]

Nonetheless, ecclesial ethics and liberal theory are not incompatible; the communitarian concern for social attachments and shared goods need not be at odds with liberalism.[59] An emphasis on community does not, for example for Fergusson, require that the individual make a commitment to relativism in morality or epistemology. The dangers of atomism – understood as the evasion of all social roles – are not identical with the dangers of liberalism – understood as free participation. For Fergusson the 'liberal ideal is not the evasion of all social roles but the protection of the space within which one can endorse, revise or even reject the commitments with which one finds oneself. The importance of communities is not neglected here. In a sense it is affirmed'.[60]

The principles of the 'common good' and 'subsidiarity', which are part of the natural law tradition and Catholic social teaching, already capture some of the insights communitarianism would like to retrieve. They are concepts that mediate not only between the individual and the state, which is of particular concern to communitarians, but also between the individual and the community. Their role in environmental ethics are discussed in the section on natural law.

Non-human creation from a Christian communitarian perspective

The retrieval of a view of the good life from a community rather than individual standpoint has much to offer environmental ethics. Indeed many communitarian themes and emphases have environmental analogues, at

---

58 Junker-Kenny, 'Virtues and the God Who Makes Everything New', 8.
59 Fergusson, 'Communitarianism and Liberalism: Towards a Convergence', 43.
60 Ibid., 45.

least in a simple sense. The emphasis, in ecology, on 'holism', 'relationality' and 'the web of life' already imply a critique of atomism. As we have seen, libertarian economistic thinking cannot deliver public goods, such as a clean environment and sustainable resource use, and this has echoes in communitarian critiques of the modern state. Many practices in environmental ethics function at the teleological level, at the level of good habits, habits learned and reinforced in community. To take a simple example domestic recycling is not obligatory, either legally or morally. Yet where it is facilitated at the community level it becomes a self-reinforcing project. Environmental values are reinforced at the level of the everyday not by immediate fear of environmental disaster or legal sanction but in the teleological vision of 'sustainable' living, where the individual freely participates for the benefit of all.

However, communitarianism in ethics has not focused on the provision of 'public goods' nor the recognition and protection of our common natural heritage. Its chief concern is with moral formation in human community and the strengthening of communal responsibility against social atomism. In its ecclesial form it expresses what a way of life proper to the church might look like. This adds another lens through which a Christian environmental ethic might be developed.

It also adds another layer to the investigation, that can distract as well as illuminate. An ecclesial approach does not immediately make the task of developing a Christian environmental ethics easier. It can obscure some of the paths that have commonly been taken on environmental questions; from creation theology to an appreciation for the value of non-human nature for example. Some communitarian approaches see a hiatus between a theological and a non-theological understanding of the world; between creation, theologically understood and nature, scientifically understood.

Stanley Hauerwas addresses the question of the value of 'nature' in Barth's theology in *With the Grain of the Universe* in the context of his ecclesial ethics. Barth, he says, shows that 'there can be no natural theology, no account of human rationality, apart from a full doctrine of God as creator and redeemer'.[61] Creation and not nature is a category for theology, according to Hauerwas. An environmental ethic, unless it refers to God, is therefore not theologically relevant. So although communitarianism might view nature as creation, it admits of no clear line from a creation perspective to an environmental one. The relationship between creation, theologically understood, and non-human nature, scientifically understood, will be explored in chapter three, which addresses environmental theologies.

Whatever the difficulties of reading from creation to non-human nature, communitarian themes have much to add in environmental ethics. It is clear that environmental problems are not easily addressed from an atomistic individualistic or libertarian economistic perspective; these approaches simply cannot register that there are public goods to be valued. Communitarianism recognises the community value of social and economic goods and explores how systematic and institutional ethics influence, even dictate, social practices or virtues. Given that such is the case it comes as no surprise that the revival of virtue ethics has had its own independent trajectory outside of explicitly communitarian and ecclesial circles in environmental theological ethics. Virtue ethics has potential in the area of institutional ethics since one of its primary themes is to investigate the continuity and change in values that support and reinforce 'virtuous' practices. These critiques register the downside of a global market economy that fails to serve marginalised communities or protect natural heritage because of inherited inbuilt lacunae and a kind of institutional inertia and myopia.

---

61   Michael S. Northcott, 'With the Grain of the Universe: The Church's Witness and Natural Theology by Stanley Hauerwas', *The Expository Times* 114, no. 4 (2003).

Summary

Communitarianism critiques anthropologies predicated on atomistic individualism. It argues for the role of community in moral formation and ethics. In its ecclesial form it can see the churches as intermediate institutions between the individual and the state. As a teleological approach in ethics it is a necessary corrective, but not an entirely sufficient basis, for a theological ethics. In my view it is a complement to, rather than an alternative for, a deontological position.

It also carries with it some 'non-accidental attendant' dangers. Radical Orthodoxy exaggerates the distance between Christian and 'secular' social achievements and political legitimacy. The urgent task of the Church is not in fact 'to demystify the nation-state and to treat it like the telephone company'.[62] Civic society is more than mutual tolerance and forbearance in the face of a plurality of mutually incompatible positions. It is ironic that radical ecclesial communitarianism undermines the very thing it claims to defend, a version of the common good and the value of 'participatory communities'. Civic society and participative democracy also hold out the possibility of developing shared understandings of duty, of justice, of the good life and of the human condition. Ecclesial communities both reflect and have helped to construct these possibilities.

Communitarianism has taken divergent forms in its critique of the dangers of individualism. The moral philosophy of MacIntyre and the ecclesial communitarianism of Hauerwas take issue with the negative effects of modern individualism on social and ethical cohesion. Radical orthodoxy extends its critique of individualism to the whole project of modern theology. Charles Taylor, however, is more modernity-friendly and specifically critiques atomistic individualism in order to develop an 'anthropology of situated freedom'. This movement has prompted a growing interest in the ethics of tradition – which includes virtue ethics, the ethics of good counsel and of the good life (*eudaimonia*) – an interest in

---

62 William Cavanaugh, 'Killing for the Telephone Company: Why the Nation-State Is Not the Keeper of the Common Good', *Modern Theology* 20, no. 2 (2004), 266.

areas of ethics said to be neglected since Kant. Celia Deane-Drummond takes up the task of applying a virtue ethics approach to issues in environmental ethics and this will be explored in chapter four. She makes a case for a virtue ethics that could foster and sustain the formation of 'habits of the heart' that are more in keeping with a Christian understanding of non-human nature as creation, by suggesting that moral formation in community can foster both 'environmental virtues' and the institutional structures that would support them. In this sense her position mediates between themes in ecclesial communitarianism and in the public domain.

Natural law

Natural law in Christian moral theology is a realist ethic that gives full status to the order of creation and assumes that we can know about the way we are to live from our human rational nature. Natural law, at least in its classical formulation, assigns to human rational nature an essence that acts as a moral guide, once correctly understood. It self-consciously develops a tradition of thought stretching back through Thomas Aquinas to Aristotle.[63] Its chief antecedent, the philosophy of Aristotle, emphasised the teleological nature of the human person – in particular the notion that human beings have a fixed rational nature and moral actions are those that fulfil one's nature as a rational human being.[64] Natural law approaches share the central tenet that moral duties can be ascertained by reflection on rational human nature, even if ancient, medieval, neo-Thomist and modern revised positions differ on what that nature is and how it can be

---

63 Gerard Bradley, 'Moral Truth, the Common Good and Judicial Review', in *Catholicism, Liberalism and Communitarianism*, ed. K. Grasso, G. Bradley, and R. Hunt (Lanham, MD: Rowman and Littlefield, 1995), 122.
64 R. Spinello, 'Natural Law', in *International Encyclopedia of Ethics*, ed. John K. Roth (London: FD, 1995), 599.

discovered.⁶⁵ Nonetheless all types of natural law claim the existence of an 'objective' moral order. As a consequence, knowledge of moral value and obligation (moral truths) is, theoretically, universally available to all 'people of goodwill' irrespective of religious commitment.⁶⁶

Revised natural law positions claim Thomas as a source but do not necessarily tie themselves, as the neo-Thomist approach did, to the classicist world view. They do not emphasise essential and eternal truths about human nature but stress the greater role of interpretation in ethical judgement. Revised positions have less of the security of the classical approach since they involve the integration of sources and the need to weigh and to recognise what it is that is ethically relevant. Classical natural law claimed *revelation* through scriptures and the *tradition* of the faith community as its sources and, for Roman Catholics, the teaching office (*Magisterium*) of the church as the interpretative authority. In Lisa Sowle Cahill's revised natural law position, there are two additional and complementary sources for natural law beyond scripture and tradition; 'philosophical accounts of essential or ideal humanity ("normative" accounts of the human); and descriptions of what actually is and has been the case in human lives and societies (descriptive accounts of the human)'.⁶⁷ These two are included on the basis that the natural and social sciences are continually bringing new knowledge to light that informs moral judgement. The philosophical accounts of the normatively human are independent of the empirical sources. So revised positions claim more sources; scripture, tradition, the normative and the descriptively human. They also claim a different authority. The authority that mediates these four is reason, as it was with Thomas, and not the Magisterium of the church.

The natural law position in ethics offers a realist ethic that emphasises the embodied aspects of human rationality without equating that with the whole of morality. It is in the concepts of the 'common good'

---

65 G. Hughes, 'Natural Law', in *A New Dictionary of Christian Ethics*, ed. James F. Childress and J. MacQuarrie (London: SCM, 1987).
66 Richard Gula, 'Natural Law Today', in *Readings in Moral Theology Number 7*, ed. Charles Curran and Richard McCormick (New York: Paulist Press, 1991).
67 Lisa Sowle Cahill, *Between the Sexes* (Philadelphia: Fortress, 1985), 4.

and its corollary 'subsidiarity' that natural law in particular has potential in working out what might constitute a 'right relationship' between the human person and the wider ecological continuum. Natural law has an implicit 'anthropocentric' focus and does not deviate from the view that non-human nature serves the human good. However, this human-centred reading cannot be interpreted as simply instrumentalist in relation to non-human nature. Rather the concepts of the common good and of subsidiarity reflect an anthropology that gives nature greater attention as part of God's creation.

Definition and history

Natural law is classically defined as 'the sum of the moral demands which can be known by reason and which are logically independent of revelation'.[68] Natural law theory has its roots in classical Greek and Roman philosophy. 'The early fathers received it mainly from the Stoic tradition (witness the frequent quotations from Seneca and Cicero) and conceived it as the order of the universe, perceived by human reason, which participates in the logos that penetrates the whole reality. It is objective and universal.'[69] In the medieval period Aquinas, also drawing on Aristotle and the Roman jurists, conceived of a 'more systematic and flexible concept of natural law, which has remained the basis of Roman Catholic ethics.'[70]

It would be a mistake to equate even classical natural law with a 'crass objectivism'. It has a teleological dimension in all its ethical questions. This is understood more in terms of the unfolding of possibilities rather than some kind of automatic passage to a predetermined goal. So in spite of its reliance on a classical, metaphysical view of human nature, natural law is still open to the possibility of change and historicity.

68 MacNamara, *Faith and Ethics: Recent Roman Catholicism*, 47.
69 J.M. Bonino, 'Natural Law', in *Dictionary of the Ecumenical Movement*, ed. N. Lossky et al. (Geneva: World Council of Churches, 1991), 713.
70 Ibid.

While in its Christian formulation by Aquinas natural law ethics depended on an Aristotelian realist ontology, the role of reason is more than just receptive to given physical structures. In addition for Aquinas, theological requirements are in accord with the demands of reason and are not contradictory. 'Neo-Thomism' continued to base natural law morality on unaided natural reason but tended to emphasis the physicalist aspects of the classicist worldview.[71] This physicalism has been criticised for being based on a naïve realism that exaggerated the importance of a static view of human biological nature in determining reality.[72]

Revised natural law, in contrast, belongs to a worldview that takes seriously experience, history, change and development. In their struggle with the perceived physicalism of the Magisterium revised natural law positions stress that the 'natural' in natural law theory should not be thought of as the opposite of 'artificial'. If it were, this would leave little room for human freedom, since 'moral knowledge would be simply a matter of discovering the given patterns in the world, and freedom would be reduced to a matter of abiding by or violating what is given'.[73] This is often how classical natural law has been (mis)construed. Revised positions defend classical positions from such a misunderstanding; natural law makes no claim to 'unjustified inference'[74] from facts to values (the naturalistic fallacy) in either its traditional or revised forms. Revised positions therefore see no contradiction, in terms of the tradition, in uncoupling natural law ethics from both an 'outdated' classical metaphysics and from the 'physicalism' of neo-Thomism.

This has been shown by Richard Gula, for whom Roman Catholic natural law is analysed as over-reliant, particularly in the area of sexual ethics, on a 'physicalist' understanding of human nature. In assessing the moral evaluations put forward by the Roman Catholic Magisterium he investigates how statements in ethics are founded; whether they are founded on the basis of the 'order of nature' or on the 'order of reason'.

71   Gula, 'Natural Law Today', 374.
72   Ibid., 373.
73   Ibid., 379.
74   Rooney, *'Divine Commands'*, 118.

In sexual ethics, he found, statements were based on 'the order of nature' and the emphasis was 'biological', whereas statements in social ethics were more 'rational' in their emphasis, being based on the 'order of reason'.[75] It may be that arguments based on the 'order of nature' have the advantage of giving a degree of certainty, consistency and precision. Those based on the 'order of reason', in contrast, can only provide more modest, cautious statements in social ethics. The danger in a shift from the 'order of reason' to the 'order of nature' is that biological human nature becomes the measure for rational human nature. This shift to a 'biological' reading undermines such evaluations, and leaves them open to the charge of being subject to the naturalistic fallacy.[76]

Natural law thinking has the advantage for ethics in that it keeps the 'embodied givenness' of human nature to the fore. However, as Gula says, the 'biological' cannot be a substitute for the moral. In other words, biological nature is not the whole of human nature and it is not possible to derive moral imperatives from bodily structures. So natural law ethics is not a naïve realism, nor does it rely on some disembodied notion of 'pure reason'. Natural law ethics does not sever the connection between physical structures and interpretations of those structures. Rather its 'physicalism' can recognise the 'givenness' of human nature and its understanding of reason makes a case for the reconstructive aspects of human knowledge and morality:

> Physicalism, however, does light up some truth. Part of being human is to have a body whose structure and functions cannot be arbitrarily treated. So the strengths of the physicalist approach to natural law is that it clearly recognises the 'givenness' of human nature. ... The weakness of this approach, however, is to mistake the 'givens' of human nature as the whole of human nature, or to take the fixed character of human existence as being closed and beyond the control of human

75 Gula, 'Natural Law Today', 369–89.
76 Other shifts have been observed in the ethical approach taken by the Magisterium. Fergusson cites *Veritatis Splendor*, the 1993 Papal Encyclical on moral theory, as an example of a move away from positions based on either the order of reason or nature to an emphasis on church community and the teaching authority of the Magisterium, Fergusson, *Community, Liberalism and Christian Ethics*, 46.

creative development. The danger of physicalism is to derive moral imperatives from bodily structure and functions and to exclude the totality of the person and his or her relational context in making a moral assessment.[77]

Natural law therefore, rightly, registers the 'givenness' of our biology but does not equate it with the whole of human reality.

It likewise gives a place to human reason. It is in its 'second strain', the 'order of reason' that the rational aspect of natural law is developed. It emphasises that the human person is not subject to their biology as if it is a matter of fate. For Gula revised natural law taps into the second strain of interpretation – the order of reason – in response to an overemphasised 'physicalism' particularly in sexual ethics. Gula provides us with a revised profile for natural law; one which is inductive, experiential, sensitive to ambiguity and open to complexity and to historical change (374).

Taking Gula's characterisation as a starting point, the revised strain of natural law has several advantages for an environmental ethics. It remains firmly a realist ethics which implies that it is open to the findings of other disciplines as sources for theology. It is teleological and historical and can thereby account for change as well as continuity in values. In common with classical natural law, revised natural law emphasises the 'personal' aspects of morality. 'Person' in Gula's reading refers to a moral subject, embodied, historical, material, related, social, called to know and worship God, unique yet fundamentally equal. In these aspects of its anthropology, the autonomy position in Christian ethics shares a common heritage with natural law morality.

Scripture as a source in natural law ethics

Natural law is a respected source for Christian ethics because it assumes that an objective moral order exists, that morality is grounded in reality, (which is not situation dependent), that it is not voluntarism, nor is it legal positivism. Natural law is also traditionally thought to be universalisable,

---

77 Gula, 'Natural Law Today', 372.

and thereby accessible independently of religious commitment, giving it potential in the public sphere. However, in terms of Christian distinctiveness, natural law approaches have been accused of being a 'minimalist mixture of philosophy and jurisprudence' and as such they could not be a sufficient basis for Christian morality.[78]

Vincent MacNamara points out that when the neo-Scholastics referred to the bible they tended to 'cite individual texts as corroborative proof of a position arrived at through natural law or taught by the Magisterium of the church'.[79] However that did not mean that their approach was 'only' a philosophical one. It was, at the same time, 'axiomatic for the neo-Scholastics that God is necessary to morality'. There is a theistic assumption in natural law which is not always made explicit, in particular by revised proponents. For Aquinas, to whom contemporary understandings refer, natural law in itself is, of course, not a sufficient basis for morality, it is sufficient only in so far as it participates in the eternal law of God, since God is the author of all goodness.

In twentieth century renewal movements the specifically biblical content of the Christian ethos of natural law was re-emphasised in Roman Catholic morality. One outcome of this renewal was a move away from natural law to an increased emphasis on the role of scripture in ethical discourse (the 'faith ethic' approach).[80] One of the pastoral tasks of the Second Vatican Council was to open the Scriptures to the people of God. The Bible was to be the soul of theology, no longer to be treated as a mere collection of 'proof texts' – as the Roman Catholic appropriations of the text were often considered to be.[81] The initial impetus for this renewal may have waned, as MacNamara suggests, but the question

---

78   Vincent MacNamara, 'The Distinctiveness of Christian Morality', in *Christian Ethics*, ed. Bernard Hoose (London: Casells, 1998), 157.
79   MacNamara, *Faith and Ethics: Recent Roman Catholicism*, 13.
80   The other strand of the renewal movement is the autonomy position.
81   MacNamara, *Faith and Ethics: Recent Roman Catholicism*, 97.

of how Christian the natural law approach is, and what distinguishes it from secular approaches, remains.[82]

It has not always been clear how philosophy and theology are linked in natural law approaches, there is no internal connection. The Roman Catholic tradition could be critiqued for only linking reason and revelation institutionally through the teaching office (*Magisterium*) of the Roman Catholic church.[83] Revised positions give more room to scripture as a source for morality than classical natural law did, but it may still only rank as one source. What they have revised is the question of authority; switching from the Magisterium to reason. Natural law revisionists, as Lisa Cahill tells us, emphasised that magisterial thinking must be accountable to criteria of evidence and logic. They sought a role for dissent, in that it must be seen not always as a threat but as a dimension of growth.[84]

The mid-twentieth century renewal movement that sought to make morality 'biblical through and through'[85] faced a counter reaction in Roman Catholicism. This counter-reaction, the early autonomy approach,

---

82 For Schockenhoff, 'the arguments adduced to justify natural-law commandments vary between metaphysical affirmations about the essence of human nature, the affirmations of a necessary unity of the order of creation and the order of redemption, and (derived from this) the competence of the church's Magisterium to provide authoritative exposition' (29). Theologians sought both to demonstrate the binding character of natural law to confirm church teaching and on the other hand appealed to the church's Magisterium when the demonstration was not convincing. 'The results have been a strange kind of positivism and an ecclesiastical moral doctrine that is both deductive and authoritative.' This undermines, for Schockenhoff, the concept of natural law as that which provides a rational justification for the fundamental affirmations of church teaching with at least the possibility of universal validity. Eberhard Schockenhoff, *Natural Law and Human Dignity; Universal Ethics in an Historical World*, trans. Brian McNeil (Washington D.C.: The Catholic University of America Press, 2003), 30.
83 Cf. Dietmar Mieth, 'Autonomy of Ethics – Neutrality of the Gospel?', *Concilium: Is Being Human a Criterion of Being Christian?* 5, no. 18 (1982), 33.
84 Lisa Sowle Cahill, 'On Richard Mccormick: Reason and Faith in Post-Vatican II Catholic Ethics', in *Theological Voices in Medical Ethics*, ed. Allen Verhey and Stephen Lammers (Michigan: Eerdmans, 1993), 84.
85 MacNamara, *Faith and Ethics: Recent Roman Catholicism*, 15.

questioned the uncritical appeal to biblical morality and returned to the methodological issue of the justification of moral obligation.[86] It still challenges revised natural law approaches to distinguish between the source of moral obligation and the content of the moral claims they make. Divine command moralists can legitimately critique natural law on the grounds that it trades in 'unconfessed theological contraband'. Likewise from an autonomy perspective a Christian reading of natural law has to make clear the distinctively Christian and philosophical premises it relies upon.

Deontology and teleology in the natural law

For Lisska, Aquinas's theory of natural law is not 'the standard paradigm of a teleological theory', he is not a teleologist in the utilitarian or consequentialist sense.[87] The moral status of an action in natural law does not depend only on the effects. If moral action is directed towards an end it is not because that end is most practicable but because it is essentially built in to the very nature of a human person. For Lisska, Aquinas would be better described as a deontologist rather than a teleologist, because his ethic is based on ends that are determined by human nature, which in turn point to a prior duty and obligation regardless of consequences. Natural law is both deontological, at least in its classical form, and teleological.

However, it is not always clear how these two aspects have been balanced in natural law theory. Jean Porter restates the deontological aspects of natural law, at least as it is found in Thomas, by reconnecting it to its roots in biblical exegesis. The deontological aspect of natural law is therefore predicated, for her, on divine command. There is, she suggests, a uniquely Christian deontology. Bradley in contrast suggests that it is the teleological aspects of natural law that are distinctively Christian and that these relativise all universal deontological claims. So the question

---

86 Ibid.
87 Cf. Lisska, *Aquinas's Natural Law*, 100.

of how these two are balanced also addresses the question of what is a specifically Christian ethic.

Specifically, Porter says, scripture privileges certain aspects of human nature. It distinguishes between aspects of our rational nature that are normative and those that are not.[88] Human nature, she rightly says, 'underdetermines' our morality, that is we are not bound or determined by natural temporal causes. In addition, Christian morality is more than a natural (rational) morality. In this way she qualifies the presumption that because of (created) reason there are ethical commonalties between 'all people of goodwill'. This does not, of course, prevent dialogue on the commonalties; on the shared inclinations and tendencies of humanity that are realised in different ways. At the same time, she says, Christian moral imperatives are not just the same as those rationally attained by others of good will.

Natural law, with its inbuilt teleology and in light of its theological and philosophical antecedents, is assumed to be well placed to engage in the (philosophical) revival of virtue ethics now underway. Yet it is unclear how the new manifestations of virtue ethics are to be related. It seems that, for some Catholic theologians virtue ethics – as an ethics of tradition – relativises the universalist claims of natural law. Bradley for example suggests that virtue ethics 'may be distinguished from natural law theory at an operational level: that is, virtue ethics moderates (in extreme examples, rejects) natural law's subject matter – choices and actions – in favour of habits, or what kind of life to live, or what kind of person to be'.[89] For Bradley virtue ethics 'questions (or rejects) the universal knowability claimed for basic principles of natural law'.[90] So for Jean Porter the distinctly Christian aspect of morality is deontological and scriptural, for Bradley it is teleological and communitarian.

Natural law is both deontological and teleological and so has historically had to deal with the question of how the level of obligation can

---

88 Jean Porter, *Nature as Reason: A Thomistic Theory of the Natural Law* (Grand Rapids: Eerdmans, 2005), 165.
89 Bradley, 'Moral Truth, the Common Good and Judicial Review', 122.
90 Ibid.

be supported at the level of values and vice versa. What is significant for this study is that natural law keeps the 'natural' in view, the embodied rational person. It does this more that other positions and so it is not surprising that it finds favour in theological ethics from otherwise hostile quarters, in the environmental theology of Michael Northcott for example. Undoubtedly one of the significant contributions it has made is in social policy, to which I now turn.

The common good and subsidiarity as contributions to a shared public moral discourse

It is in its commitment to a shared moral discourse that the strength of natural law lies. In contrast to the ambiguous situation we are left in with communitarianism, the natural law tradition protects the idea that certain fundamental values can be shared by diverse peoples.[91] Natural law ethics emphasises the dignity of every human person as created and evaluates the moral and cognitive capacity of human reason positively. Although it defends the individual, it is not individualistic. In natural law intersubjectivity and reflexivity both belong to (Christian) anthropology. It contributes, in the public realm, specifically under the rubric of the *common good* and *subsidiarity*.[92] Both of these concepts can mediate between the individual and the state expressly because natural law does not assume that human beings are asocial by nature nor that government is an artificial construct.[93] In order to explore these two we must first discuss the dimension of sociability which they defend.

---

91 Konrad Hilpert, 'The Theological Critique of "Autonomy"', *Concilium: The Ethics of Liberation–The Liberation of Ethics* 2, no. 172 (1984), 13.
92 The 'good' implies the teleological (perfective) notion of 'the good' characteristic of traditional Thomistic theory and presupposes a particular understanding of nature, especially human nature.
93 Grasso, Bradley, and Hunt, eds, *Catholicism, Liberalism and Communitarianism*, 92.

Sociability

Politically, natural law, almost by definition, is compatible with limited government, rather than any kind of theocracy. It implies 'the existence of a whole area of life that is remote from the competence of government.'[94] This whole area of life is not simply centred on the individual, rather sociality is viewed as internal in relation to persons. And this natural sociality is not understood in contractual terms. For Grasso et al., indeed, the Catholic 'human rights revolution' can combine the truths of liberalism and communitarianism while avoiding the errors of either.[95] The communitarian critique of atomistic individualism has contributed to a confident restatement of the principles of Catholic social teaching. These concepts – the common good and subsidiarity – are two concepts which support the fundamental principle of the dignity of the human person in Catholic social thought and at the same time assuage the concerns of communitarians about atomistic individualism.[96] This is because they are:

> ... not rooted in a political theory that assumes that human beings are asocial by nature and that government is an artificial construct, a phenomenon made necessary by the defects of man – somewhat regrettable, but necessary ... It sees man as a naturally political being, who achieves his good through natural communities (political as well as familial) and not just voluntary contractual arrangements.[97]

This natural law view of the public realm contrasts sharply with divine command and communitarian approaches. This is not strictly because it sees sociality as internal to a Christian anthropology – those positions would share that conviction. Rather it differs from them in that natural law does not view this sociality as artificial or external to non-Christian society. The secular realm is not characterised by 'ontological violence' as

---

94   Ibid., 9.
95   Ibid., 11.
96   Cf. Christopher Wolfe, 'Subsidiarity: The "Other" Ground of Limited Government', in *Catholicism, Liberalism and Communitarianism*, ed. K. Grasso, G. Bradley, and R. Hunt (Lanham, MD: Rowman and Littlefield, 1995).
97   Ibid., 92.

the radical orthodox position would have it. On the other hand, natural law ethics would take for granted to some extent, that the secular realm is indebted to the Christian social project, even if it might not want to admit it, as O'Donovan has reminded us.

The common good

The concept of the 'common good' captures how natural law ethics envisages the relationship between the individual and the community. At the centre of the tradition is a conception of justice that involves the rights and duties of all members of society and their interdependence in the pursuit of the common good, in the responsibilities of the state and in the legitimate sphere of independence of local groups and institutions (subsidiarity).[98] The natural law concept of the common good aims ideally to include 'elements which individualism is both unable to account for in theory and likely to neglect in practice'.[99] Social justice is tied up with individual freedoms but it is also served where institutions and structures mediate the claims of individuals.[100]

Undoubtedly the use of the concept of the 'common good' has not always been exemplary. It has, for example, relied too heavily on the 'doctrine' of the 'trickle down effect' in relation to distributive justice.[101] As Cahill points out 'one endemic shortcoming of the past Catholic *common good* tradition is that, while it has increasingly turned attention to the plight of the poor, it has usually still conceived of social change as ema-

---

98   Cf. Sowle Cahill, 'Genetics, Individualism, and the Common Good', 387.
99   Ibid., 388.
100 In Finnis's view the common good cannot be simply equated with the public good, it refers to more than the political good. Cf. John Finnis, 'Public Good: The Specifically Political Common Good in Aquinas', in *Natural Law and Moral Inquiry: Ethics, Metaphysics and Politics in the Work of Germain Grisez*, ed. Robert P. George (Washington: Georgetown University Press, 1998).
101 It is only since the 1990s that economists began to question the efficacy of the 'trickle down effect', even when there were genuine attempts to find a path to distributive justice and equity in economic analysis, as Amartya Sen has demonstrated in his work.

nating from the top down, and has directed its call for change primarily at those enjoying power'.[102]

Read as a normative teleological approach in ethics, the concept of the common good does appear to lack the advantages of a deontology. Lisa Cahill, however, would defend its suitability as a social principle. She says that the concept of the 'common good' is not a simply a question of 'balance' between the rights of the person and that of the community.[103] Rather the common good has heuristic potential for addressing problematic situations in social and personal ethics. Her heuristic understanding of the common good, which is in keeping with the tradition, is also compatible with an autonomy approach which seeks to integrate the intersubjective aspects of personal autonomy and dignity. This use of the common good concept, to discover rather than play off the good of the individual against that of the community, has heuristic potential for environmental ethics. However, from an autonomy perspective it remains in need of a deontological framework, so that the dignity of the human person is not sacrificed to a communal environmental goal. Part of its contribution in environmental ethics is in its sister concept of subsidiarity.

Subsidiarity

The principle of subsidiarity is a criterion, rooted in the common good, for discovering ways to integrate the individual and the social good. It also safeguards the mediating institutions of civil society, between the state and the individual.[104] It was first articulated in *Quadragesimo Anno* (1931) by Pope Pius XI in the following terms:

> Just as it is gravely wrong to take from individuals what they can accomplish by their own initiative and industry and give it to the community, so also it is an injustice and at the same time a grave evil and disturbance of right order to assign

---

102 Sowle Cahill, 'Genetics, Individualism, and the Common Good', 389.
103 The notion of the 'public good', in contrast to the common good, sets 'balance and harmonisation' both as an ultimate goal and a pragmatic principle.
104 Elshtain, 'Catholic Social Thought, the City and Liberal America', 98.

to a greater or higher association what lesser and subordinate organisations can do. For every social activity ought of its very nature to furnish help to the members of the body social, and never destroy or absorb them.[105]

This implies that there should be a dispersal of authority to the local level, at least to the degree that is compatible with solidarity and good governance.

However there is another, equally significant, inference that can be drawn from the concept of subsidiarity. It would be a misreading of it, given that its corollary is indeed the common good, to imagine that it privileges one 'locality' over another. The task for ethics is to identify what the 'local' refers to. Issues in institutional ethics cannot be addressed effectively at the level of the individual or perhaps even at the regional or indeed national level. Subsidiarity is a criterion which can help to identify the appropriate 'operational level' in ethical deliberation. Many environmental problems, for example, can only be addressed at the global level, and consequently the global must be understood as the most local level.

In the arena of economics the vexed question of debt relief serves to illustrate this heuristic aspect of subsidiarity. Sabine Alkire, a natural law ethicist, questioned whether or not the World Bank Group had a responsibility for debt relief. At first glance it would seem that debt relief is neither the task nor the objective of the bank, given the purposes for which that institution was established.[106] However, applying the criterion of subsidiarity Alkire 'discovers' that the responsibility for debt relief in fact 'bubbles up' to international institutions, such as the World Bank and the International Monetary Fund (IMF). Debt relief, she says, cannot be tackled by individuals, civil society or nations states; this is not because the sums are large but because the architecture of debt is 'thoroughly

---

105 Catholic Bishops' Conference of England and Wales, 'The Common Good and the Catholic Church's Social Teaching' (London, 1996), 51.
106 Sabine Alkire, 'When Responsibilities Conflict: A Natural Law Analysis of Debt Forgiveness, Poverty Reduction, and Economic Stability', *Studies in Christian Ethics* 14, no. 1 (2001), 72.

international'. So the World Bank and the IMF have the 'unique ability' to tackle this problem. In addition, she continues, because they have this unique technical and political ability to convene debt relief they also have a '*unique responsibility to do so*'. Or they have at least the unique responsibility to co-operate with other agencies in this regard (73). Admittedly this responsibility has to be integrated with other responsibilities such as securing economic stability in the long-term or ensuring that debt relief meets the needs of those it sets out to assist (72). It also requires the determination of what alternatives actually exist (76) and the time frame in which they can realistically take shape (75). She acknowledges that subsidiarity does not adjudicate between these aspects of the problem; between conflicting schools of thought on possible best outcomes. Yet, it is no small thing to say that the principle of subsidiarity requires that an institution acknowledge the power and responsibility that it has, and that it must communicate this effectively. In the case of debt relief this at least, she says, avoids the damage of over-stated expectations and gives a clear baseline from which a more just institution could emerge. For Alkire the 'essential insight of this principle is that maximal responsibility for temporal affairs should be assumed by the most local organizations capable of carrying them out' (73).

In summary the concept of subsidiarity does not pitch the local against the global by assuming that institutions at the very local level automatically serve the common good more fully. Rather it suggests that the concept of the common good recognises intermediate institutions at different levels; local, regional, national, international and global. Subsidiarity is a criterion for identifying the level at which efforts might be directed in the search for more just structures and institutions that serve rather than enslave the human person in society.

Non-human creation and rational human nature

Natural law ethics is realist and anthropocentric. It has a long history of integrating the normative and empirical aspects of the human rational nature. It recognises the autonomy of created structures and the fact that

Christian ethics has a social, political and universal dimensions to its personal message. It is in these aspects that we can consider the question of how it might yield an environmental ethic or an 'ethic of production'.

Natural law stands in a tradition with a long and subtle history of debate on the moral significance of the natural. The focus in natural law is not however on the non-human creation but is firmly on human rational nature. As a system of ethics natural law is characterised by a double perspective in that it can 'refer to the ideal rational nature of the human person or to his empirical structure as a natural being with needs'.[107] Classic natural law can be characterised as chiefly deductive and revised natural law as inductive. Jean Porter rejects much of the inductive reconstruction of natural law in its revised forms – in particular the attempts by Finnis and Grisez to develop substantive notions of the good. Of course she does not deny that the 'natural' may have normative force, just that it does not 'determine' what it means to be human. Nevertheless she does ask how the Thomistic tradition might expand to address the urgent question of the relationship between humanity, individually and collectively, and the natural world on which we depend.[108] She does not doubt that it would involve a reassertion of the legitimacy of humanity's use of creation. However, she also points out that it would indicate several other things at the same time. It would suggest that any such use would be directed towards the good of all humanity and not just one society. Secondly, since Thomas assumed that all creatures possess an intrinsic goodness apart from their usefulness to humans – in so far as they participate in the goodness of God – all creatures deserve consideration; at least to the extent that a distinction be made between legitimate use and wanton use (178).

There seems little doubt that Aquinas followed Augustine in arguing that animals and plants are fundamentally ordered for human ends and that humans may even be cruel to them in their quest for life's necessities. Indeed it would be difficult to argue that an environmental ethics based on

---

107 Schockenhoff, *Natural Law and Human Dignity*, 20.
108 Cf. Porter, *The Recovery of Virtue*, 178.

reading Thomas Aquinas would be anything other than anthropocentric, according to Benzoni. For Thomas, nonhuman creatures are instruments to the human good.[109] Can a reorientation towards God assist in developing a less instrumentalist reading? For Benzoni even this 'seeing creatures as God sees them means seeing them as instruments to the human good' since God created them for human use (458). This may not seem like a good place to begin for a Christian environmental ethic. After all, the good of other creatures can only be secured in this way if their good coincides with the human good (474). For Benzoni, it seems that Thomas did not hold to such a coincidence; for him the final good of humans and other creatures diverge irreconcilably. Yet Benzoni has set out to rehabilitate Thomas for a noninstrumentalist environmental ethics. He does this by securing a noninstrumentalist value for creatures in the divine intention, so that from the 'divine' perspective all creatures contribute to the 'divine good' and therefore have non-instrumental value.

The perennial presumption in theological environmental ethics is that anthropocentrism is always to be overcome, including in Thomas. The need to 'rehabilitate' Thomas is echoed in the work of Northcott and Deane-Drummond. The only solution to the instrumentalisation of the non-human world is to rely heavily on a theocentric realignment as the one place where it is possible to secure an 'intrinsic' or at least noninstrumentalist value for the non-human creation. Michael Northcott reads the natural law tradition as a fundamentally theocentric ethic, one which has been distorted in its neo-Thomist and revisionist approaches. Like Benzoni, he says, in order to discover the moral significance of the natural, we must turn 'back to God, not to nature' as the only adequate response to the environmental 'crisis'.[110] However, insisting that a non-instrumental value for non-human creation can only be secured before God, in his goodness or intention, leaves humanity off the hook too easily. The task for Christian ethics is to discover what liberation for the non-

---

109 Francisco Benzoni, 'Thomas Aquinas and Environmental Ethics: A Reconstruction of Providence and Salvation', *The Journal of Religion* 83, no. 3 (2005).
110 Michael S. Northcott, 'Ecology and Christian Ethics', in *Cambridge Companion to Ethics* (Cambridge: Cambridge University Press, 2001), 224.

human creation, in relation to humanity, might look like in 'advance' of its promised ultimate salvation.

From a different perspective, Lisa Sowle Cahill attempts a universalist political project in working out the implications of Catholic social teaching as rooted in natural law ethics. Her emphasis has significant implications for the transformation of structures to deliver an environmental ethics. Cahill specifically sees potential in the concept of the common good for a working out of a 'right relationship' between the natural sciences and technology – albeit in light of the Human Genome Project rather than environmental questions. Cahill suggests that religious voices may function to encourage sensitivity and an interest in the 'common good' beyond minimal requirements. 'In particular they may function to encourage public sensibility beyond the minimal moral duties required by the principles of equality and autonomy, evoking the altruism and willingness to sacrifice self-interest'.[111] It is frequently these sensibilities that are called upon to instigate more sustainable uses of natural resources. Cahill's emphasis is likewise compatible with the autonomy approach in Christian ethics in that it seeks to rehabilitate and reintegrate an ethics of tradition and of the good life with a normative and universalist project.

Summary

Natural law has a universalist orientation and is a realist ethic that gives full consideration to the order of creation. In its revised form it focuses on the concept of human flourishing rather than on the teaching office of the church, or a distinct ecclesial ethos. Reason is the authority that weighs the sources for revised natural law approaches. Although scripture is a source for natural law ethics, in all its forms, natural law does not clearly distinguish between the source of moral obligation and the content of moral claims in Christian ethics. Natural law is both deontological and teleological. It has a strong theology of the public realm,

---

[111] Sowle Cahill, 'Genetics, Individualism, and the Common Good', 387.

worked out in the concepts of the common good and subsidiarity. In environmental ethics it is, at least traditionally, anthropocentric, but not necessarily instrumentalist. Re-readings of the tradition have sought to provide a non-instrumental value for the non-human creation by suggesting a theocentric shift back to God. From an autonomy perspective this lets humanity off too easily; only in God can the non-instrumental good of the non-human creation be secured.

The fourth and last position I will investigate in theological ethics, the position which this book defends, is the autonomy approach in Christian ethics. It has its roots in natural law, beginning with Alfons Auer, but it cannot be accommodated within it, as we shall see from Dietmar Mieth's critique. The autonomy approach may be seen as a 'legitimate continuation of the natural law in the circumstances created by the modern consciousness of freedom' or as a 'late representative of the classical Catholic natural law ethics'.[112] However, the Kantian notion of self-legislation cannot be accommodated in the classical form of natural law. Like natural law it is a realist ethics (transcendental idealism), with a universalist orientation and a share in the integration of deontological and teleological approaches. It differs from classical natural law in that the good is not based on a given human nature but is a reflection of historical experience, a shift that it shares with revised natural law. Also in common with revised natural law, the autonomy approach gives a greater role to human self-determination and recognises that there are a plurality of forms of the good life.

However, an autonomy approach insists that the concepts of dignity, autonomy and freedom have a greater capacity to integrate philosophical and theological insights in ethics than the concept of human nature (reason) either classically understood or in revised form. The former is tied to an outdated metaphysics, the latter to a teleological, substantive notion of the good. The autonomy approach insists that teleological ethics need a Kantian deontological framework to prevent it from being co-opted by instrumentalist readings. The categorical imperative secures the individual from the violation of others and violation in the name of

---

112 Schockenhoff, *Natural Law and Human Dignity*, 2.

the collective. In Kantian autonomy the concept of human dignity and freedom contains more than questions of morality, it also points to the question of meaning. In its Christian rereading it insists that God's self-disclosure in Jesus Christ enables people both to experience God and to realise their moral commitment through this freedom.

Autonomy in Christian ethics

Autonomy in a Christian context or the Christian autonomist position emphasises human freedom, rather than rational nature as such, as the basis of human dignity. This freedom is understood as the capability for moral self-government. In common with natural law theory it plumbs for ethical norms that can be universalised. In its theological reworking it also seeks to develop social notions of the good. What are considered to be specifically Christian in this position are the intensifying, motivational, heuristic, integrating and relativising dimensions of a Christian faith background rather than the 'content' of ethical obligations, although these can influence each other. It is in its recognition of the dignity and freedom of the human person that autonomy is considered to offer the most adequate categories to express the theological message of salvation. It is also compatible with the tradition of valuing human reason and morality in Catholic theology, and with much of German-speaking Protestant theology. A specifically environmental theology, from this perspective, would suggest that the natural world that is appreciated as good teleologically, also belongs, deontologically to the concept of human dignity itself.

Definition and history

In the autonomy approach the human person can judge his or her own actions on the basis of a supreme principle of morality which Kant refers to as the *categorical imperative*. The categorical imperative expresses the idea that 'morality begins with the rejection of non-universalisable principles';[113] that we should only act according to those maxims which can be willed a universal law. In its humanistic formulation it states that we should treat each human being, always at the same time, as an end in themselves and never only as a means. Each self-determining moral agent ought to respect the equal right of the other and thus only adopt those principles that are universalisable. Far from being individualistic and atomistic, Kant's ethic is intersubjective. The freedom of the other is the limit to individual freedom; intersubjectivity is not merely a consequence of autonomy, it is internal to it. 'In the original context in Kant's practical philosophy (1787), autonomy exists ... within the frame of intersubjective obligation'.[114] The recognition of the autonomy of others as the supreme limiting condition of our own autonomy arises not as a consequence of an autonomy perspective but as part of its very self-definition. For Kant the supreme 'limiting condition' is the dignity or pricelessness of the human person which is predicated not on human reason or nature but on the capacity to be moral; on human freedom.

Reason and faith are not sources in conflict in the autonomy approach, as they can be in some divine command and communitarian approaches, even if they do constantly need to be integrated. It is at the limits of reason that the question of meaning occurs; and it finds a religious answer, even in Kant's moral philosophy. The conflicts in human experience, the tragedy, the regret and awe that disclose a properly religious sense of the finitude and frailty of the human condition[115] can be reconciled or borne in the hope that faith brings; the possibility of a resolution beyond human means. Reason itself admits to these limits. The experience of moral obligation is based on human freedom. From a theological perspective

---

113 Cf. Onora O'Neill, 'Kantian Ethics', in *A Companion to Christian Ethics*, ed. P. Singer (Oxford: Blackwell, 1991), 177.

114 Maureen Junker-Kenny, 'Embryos *in Vitro*, Personhood and Rights', in *Designing Life?: Genetics in Human Reproduction*, ed. E. Graumann and S. Hildt (Aldershot: Ashgate, 1999), 155.

115 Fergus Kerr, 'Moral Theology after Macintyre: Modern Ethics, Tragedy and Thomism', *Studies in Christian Ethics* 8, no. 5 (1995), 44.

human fulfilment is in need of a God, one who does not allow our moral efforts end in vain. We do not need God to do the right thing or to be good to each other. An autonomy position combines the insights of faith with dialogue with non-believers on moral grounds.

Not surprisingly, the autonomy position has consequently been accused of being deism in disguise, or indeed the obvious choice of the agnostic. Yet although it is the case that an autonomy approach does bracket the God question, it does so only at the level of justifying obligation, not at the level of meaning. According to Mieth, the main feature of moral autonomy is the methodological separation of Kant's two questions, *what should I do?*, and *what should I hope for?*.[116] In this approach the source of validity of moral obligation is substantially independent of religious premises. However, Mieth rethinks these two questions in his discussion of Kant and finds that it is clear that the interpretation of moral obligation leads to a question of faith.[117]

Kant demolished the rational arguments for the existence of God in his *Critique of Pure Reason*, but in his *Critique of Practical Reason* came up against what he referred to as the 'antinomy of practical reason'. For Kant the highest good for the human person is the coincidence of virtue and happiness. Yet there is no reason to expect a necessary connection between the two. It is possible then to act, as say a humanist agnostic might, morally and virtuously but without reward, despairing of the possibility of meaning. Kant however – on the practical evidence that both the moral impulse and the orientation towards happiness are part of what it is to be human – was compelled to postulate the existence of God: the author of reality who reconciles these two. The conflict in human experience – that being moral may not result in happiness – is resolved in the hope that faith brings.

> As a humanist agnostic, all one can say in the end is, 'Even without hope for the suffering humanity of the past, the present or the future, I can do nothing other than follow the logic of responsibility' ... Judged by criteria of efficacy, ethical

---

116   Mieth, 'Autonomy of Ethics – Neutrality of the Gospel?', 35.
117   Ibid.

commitment often seems absurd; it is like a tiny flickering flame in a darkened world. But the Gospel tells us that God takes responsibility for efficacy, and gives the whole world a share in the little we contribute. Because through faith we have a reason for real hope, we understand the meaning of obligation.[118]

In theological perspective, the autonomy position goes beyond the methodological separation of Kant's two questions. It is not an autonomy ethic simply overlaid with a faith ethic. It is a moral and ethical approach that in its faith context argues theologically and philosophically for an autonomous foundation for morality. What Mieth specifically wants to express is that positions based on revealed morality must make their philosophical positions explicit to avoid interpreting revelation just to suit their needs. In this way he connects with O'Donovan's project, which is to make explicit to secular culture what it owes to Christianity.

Mieth evaluates Kant's conviction – that freedom is a condition for the possibility of morality – positively, in contrast to O'Donovan's negative assessment of autonomy. Mieth sees it as compatible with the Christian faith. In the Roman Catholic tradition, the tradition in which this autonomy approach is articulated, there is already the specific recognition of the autonomy of created reality, if not the autonomy of ethics.[119] The autonomy position stresses the contribution that other disciplines bring to questions of morality, in a relationship of mutual critique. It also registers the fact that many non-Christians live out high moral ideals.

Admittedly the concept of dignity and the principle of autonomy, have been invoked in contradictory ways.[120] After World War II in Europe 'dignity' was employed to contain the power of states over persons and was central in the Declaration of Human Rights. Yet the concept of respect for the dignity of persons appears to have divided, at least stereotypically, into two schools. In the American school respect for individualistic, even

---

118 Ibid., 39.
119 Dietmar Mieth, 'Autonomy or Liberation: Two Paradigms of Christian Ethics', *Concilium: The Ethics of Liberation–The Liberation of Ethics* 2, no. 172 (1984), 88.
120 Patrick Verspieren, 'Dignity in Political and Bioethical Debates', *Concilium: The Discourse of Human Dignity* 2 (2003).

atomistic, self-determination took precedence over that of dignity.[121] In Europe human dignity retained its place alongside the concept of identity and the respect for the integrity and other fundamental rights and freedoms of the human being, even if each state can defend who is the bearer of human dignity.

The corrosive impact of atomism is not inherent in Kant's account of autonomy, as many critics of Kant would suggest. Kant's emphasis on individual moral autonomy does not imply that we have no social obligation. His 'Kingdom of Ends' is a world in which 'we first acknowledge our likeness to one another as the very condition of our being moral agents'.[122] It is inherently intersubjective and thereby social, identifying a constraint or limit on our actions in relation to others, which we may not override or control.[123] It is precisely not atomistic in its individualism. So although the principle of autonomy may be interpreted to include that an individual has 'adopted a life plan of their own reflectively and critically', Kantian autonomy is not identifiable with atomistic individualism.

Christian distinctiveness in five dimensions

For Oliver O'Donovan both natural law and autonomy have effectively disenfranchised Christian ethics. They

> ... make a virtue of denying that 'Christian ethics' in the strict sense can exist. Such theories may allow that Christian faith has a bearing on ethics indirectly, in that Christian spirituality promotes a heightened concern for the moral dimension of life and a strengthened ability to cope with it. But the substance of ethical questions, they hold, is not open to special illumination from the gospel; the believer is in no more favoured a position than the unbeliever when it comes to discerning the difference between good and evil.[124]

---

121   Ibid., 17.
122   Mieth, 'Bioethics, Biopolitics, Theology', 7.
123   Junker-Kenny, 'Valuing the Priceless', 48.
124   O'Donovan, *Resurrection and Moral Order*, 11.

O'Donovan raises here an enduring question for both approaches, and indeed for Christian ethics in general. Does Christian faith provide us with a specific content for morality? For the divine command moralist the answer is confidently in the affirmative. Of course, divine command morality is still faced with a complex task; Christian ethics has to continue to work out what this specific content is. In natural law and in autonomy the discussion has taken a different trajectory. These approaches have taken it as on ongoing task to continue to grapple with this particular question, to show how faith influences morality and in what way faith might impinge on the question of content. In this section I will briefly sketch this debate in ethics, which reflects an enduring tension in Christian theology between understanding the human person as already graced by God in creation but also in need of God's self-disclosure in specific revelation for salvation. The autonomy position in Christian ethics dealt with this question by relocating morality and ethics in the transformative 'context' of faith. This study suggests that Christian faith gives us a distinct morality in five dimension. In common with Sowle Cahill's natural law approach, this study is suggesting that the Bible is a source for Christian ethics, alongside the Christian tradition, and normative and descriptive accounts of the human. In contrast to natural law the autonomy position argues it is not reason as such, but autonomy that acts as the authority. First, however, we must turn to the question of a specific content for Christian morality.

The question of content in Christian ethics

There is a sense in which there must be a Christian morality, one that is in keeping with the Christian world view. However, suggesting that the Christian world view influences morality does not directly address the substantive question of content. The question of specific content asks whether or not there are specific moral demands made on Christians, demands that would, by implication, be unintelligible to the humanist or non-Christian? Vincent MacNamara traces the history of this debate in Roman Catholicism since the Second Vatican Council. The renewal movement, from which the autonomy position developed, he recounts, held to the view that 'morality is not to be regarded as the command of God imposed on us in revelation, ... that morality is discovered by human reason, ... that the content of Christian morality is not specific but available to us without revelation, ... that the special character or identity of Christian morality is in something other than content'.[125] Opponents of the autonomy position in contrast argue that there is behaviour demanded of the Christian that can only be justified by appeal to Christian belief. The values of poverty, renunciation of power and humility could be considered distinctively Christian (96). However, even for those who support a 'faith ethic', it is 'difficult to show that one can deduce a particular and specific moral rule from a particular theological truth' (120).

The question, of course hinges on what is meant by specific moral content. Clearly good practice can be found outside of Christianity and, indeed, morality has a certain autonomy in that 'religious faith is not necessary in order to experience and recognise the moral point of view'.[126] What then is implied by the idea of unique content? If content 'refers to the moral characterisation of an act, i.e. how the act is to be described' (107) than it is more than a issue of outward practices (examples of which can undoubtedly be found outside of Christianity, such as brotherly love, self-sacrifice or the renunciation of one's rights for the sake of the other). Content, after all, can be evaluated both in terms of its outward effects (actions) and the disposition and motivations of the agent (virtues and values). Given that that is the case, that the content of a moral act includes dispositions and motives, then in that sense, the Christian context impinges on the content of morality. The bald assumption that Christian morality does not yield any substantive content clearly needs some qualification. The autonomy position defended by Mieth and outlined here, which wishes to reconnect an ethics of obligation with that of the good life and of virtue and disposition, clearly needs to be developed further if it is to show how the question of content is specifically or uniquely influenced by a faith perspective, particularly at the level

---

125 MacNamara, *Faith and Ethics: Recent Roman Catholicism*, 50. Further page numbers are in the text.
126 MacNamara, 'The Distinctiveness of Christian Morality', 152.

of values and dispositions. As MacNamara points out, a more complex concept of moral reasoning, that understands the human person not as an impersonal rationality but as being situated in a world view influenced by religious considerations, might be required (119). He would go further than the early proponents of this position and say, as regards content, that it' is more correct to say that some of the realities of the faith enable one to see the ground of moral judgement more easily or more fully and securely' (108). This implies that a broader definition of the moral agent, which includes motivation and judgement, might be capable of elaborating on the question of uniqueness in Christian morality (114). It is a task beyond the scope, however, of this study to investigate this tension, and the precise relationship, between content and moral judgement.

However, even given that qualification on the question of content, and the need for further development, the autonomy position defended here emphasises that the Christian world-view impinges on morality, by resetting moral actions and dispositions in the context of transformative faith. It switches the question from one of content to one of context, a move which has an advantage in that the concept of 'context' is capable of situating mortality in a web of beliefs.[127] From an autonomy perspective, consequently, being Christian is a specification of being human, not an alternative (as suggested by some theological communitarians).[128] Christians have the advantage of revelation, and of a theological anthropology developed on the basis of that but, even for O'Donovan, that does not relieve them of the challenge of being moral agents. Given that qualification I now turn to the issue of a distinct Christian context.

A distinctive Christian morality and ethics in five dimensions

A Christian reading of the autonomy approach takes seriously its faith tradition's universalist claim of attributing freedom to the human person, and the context of Christian practice. It sees the theological input not as

---

127 Ibid.
128 Mieth, 'Autonomy of Ethics – Neutrality of the Gospel?', 32.

yet another 'type of ethics' but rather as 'the context' in which Christians do ethics. The emphasis on context in autonomy, rather than content, coincided with a shift in systematic theology that put the stress on God's loving us first. It suggested that rather than considering the human person as existing in two orders, that of creation and that of redemption, there is one order, that of grace. 'Current thinking recognizes that there is but one order, the order of grace, that all people exist in that order and that to everyone the one destiny of union with God is offered: it sees all people living in the order of graced nature'.[129]

The autonomy approach presented here, argues that Christianity gives a new significance and meaning to morality without changing its ground of validity, which is articulated philosophically and is publicly accessible.[130] For Junker Kenny the 'idea of God is only excluded at the level of obligation; we owe each other mutual respect because of the unconditional nature of our freedom, not because of divine command'.[131] However, the specifically human question of meaning and of hope, that arises at the limits of reason, finds a theological answer in the Gospel. This study argues that Christian faith influences morality and ethics in five specific dimensions. It intensifies our moral and ethical sensitivities. It motivates us to act. It provides heuristic potential in ethics. It integrates our sources across disciplines and lastly, it relativises morality as such. This has consequences for how Christian ethics envisages the non-human natural world in environmental ethics. In affirming that nature is the appropriate space for human life, rather than incidental to it, the realities of faith inspire and sustain our commitment to our responsibilities for all creation. They do so in five dimensions.

First, the Christian context intensifies our sensibility to our obligation and to what is good for human flourishing.[132] The biblical text, for example, contains ethical models of creatureliness that intensify our moral sensibilities. Included in these are De Tavernier's list:

---

129 MacNamara, 'The Distinctiveness of Christian Morality', 158
130 Mieth, 'Autonomy of Ethics – Neutrality of the Gospel?', 38.
131 Junker-Kenny, 'Embryos in Vitro, Personhood and Rights', 156.
132 Mieth, 'Autonomy of Ethics – Neutrality of the Gospel?', 39.

> ... love of enemy and renunciation of retribution, a critical readiness to abandon one's own legal position, the discarding of short-sighted prejudices, showing solidarity with others and having a watchful concern for the weak, the disadvantaged and the defenceless who have no legal rights.[133]

Secondly, the Christian context is relevant to motivation both for our everyday action and in relation to the question of ultimate meaning. Understanding the natural world as a gift, as the creation of a good God has moral implications for how we value non-human nature. It effects our motivation to act to transform unjust institutions and can maintain our motivation even in the face of personal failure. Faith can also provide the impulse to engage in ethical debate in the first instance. In acknowledging our imperfect historicity, a faith context can prevent ethical judgement lapsing into a naive notion of an earthly utopia, or conversely into an unwarranted pessimism.[134]

The Christian context also has heuristic potential in ethics and morality in that the 'taboos' of tradition act as 'cautionary fences' that represent historical insights or experiences and are not just artefacts from a previous unenlightened generation.[135] They represent reasoned responses to questions of God, the human person and morality in changing historical contexts. They are also living traditions of the historical experience of the good.[136] The Christian context alerts us to proposals that erode (or which may be compatible with) the dignity of the human person, philosophically and theologically understood. It also reminds us of the interpretative character of ethical evaluation. As Mieth points out, we are caught in a hermeneutical circle and we 'define its starting point because we want to act in a certain way, and we act according to how we have defined it'.[137]

---

133   Johan De Tavernier, 'Human or "Secular" History as a Medium for the History of Salvation or Its Opposite: "Outside the World There Is No Salvation"', *Concilium: No Heaven without Earth* 4 (1991), 10.
134   MacNamara, *Faith and Ethics: Recent Roman Catholicism*, 46.
135   Mieth, 'Bioethics, Biopolitics, Theology', 21.
136   Dietmar Mieth, 'Continuity and Change in Value-Orientations', *Concilium: Changing Values and Virtues* 3, no. 191 (1987).
137   Junker-Kenny, 'Embryos *in Vitro*, Personhood and Rights'.

If definitions have practical intent then theological conceptualisations can alert us to that intent.

Fourth, a Christian reading of the autonomy approach takes seriously the insights of its faith tradition and of the human sciences in the context of Christian practice, and integrates these two. Mieth develops a 'conductive method' in ethics that mediates between the normative (deductive) and contextual (inductive) aspects of moral reasoning. It demands that we both translate and critically appropriate contemporary interdisciplinary and ethical categories for Christian ethics.

Lastly, the Christian autonomy position 'relativises' moral insights. The epitome of being human, from a theological perspective is not moral capacity as it is expressed in Kant but the need for salvation, the 'feeling of absolute dependence' first described by Schleiermacher.[138] Morality is part of the human response to being a creature, with all that implies: the endowment of beauty and the spontaneity in life alongside the need for forgiveness, for redemption, and for hope in the face of the tragedy of the human condition. For Mieth the 'feature which makes the sermon on the mount and other Gospel maxims Christian is more than morality'. It is the dimension of hope that shows us that the 'Christian use of ethics is not directed to the imposition of discipline on the community or morality on the world at large, but to liberation ... Ethics too is not meant to judge, but to save.'[139]

All of these dimensions provide us with a distinctive Christian morality that influences and enlivens moral judgement. The concept of creation, and of creatureliness, as we shall see, goes far beyond questions of the valuing of 'nature' found in contemporary environmental ethics.

---

138   Mieth, 'Bioethics, Biopolitics, Theology', 18.
139   Mieth, 'Autonomy of Ethics – Neutrality of the Gospel?', 37 & 39.

Obligation through freedom

Kant's theory of practical reason is the standard-setting example of a deontological ethical theory. It defines ethics as a theory of obligation to treat the human person always, at the same time, as an end in themselves, as an autonomous person. It argues for the priority of the right over the good.[140] In its theological re-working, the autonomy position in Christian ethics, recognises that the rule theories of normative ethics are not the whole of morality and ethics. In the theological appropriation of autonomy by Mieth the ethics of tradition – virtue ethics, the ethics of good counsel, the ethics of what is good for human living – are reinstituted. Here I will examine how Mieth reintegrates the deontological and the teleological in his Christian ethics through his conductive method.

Mieth sees the inadequacy of a deductive approach in ethics that fails to contextualise its normative principles. He also critiques the inductive method because it so easily loses sight of the normative aspects of ethics. By contrast he specifies a 'conductive' method in ethical deliberation. This method follows several steps. One must first constitute the 'facts', then define the problem, and finally discuss what norms apply. All of this may seem commonplace but it is not always obvious what axiomatic framework a case falls into. Ethics, he suggests, is not just about taking a stance but is the 'science of argumentation'. Its judgements, because of this, may not be normatively obligatory, may be less sure, and may carry only a provisional legitimation. Nevertheless they are neither reductionist nor empiricist.

An application of this method can be seen in his discussion of human cloning.[141] He outlines three reasoned reactions to the possibility of the cloning of human beings and treats each of them in turn. The first involves an abstract endorsement based on a best-outcome view. This, he suggests is predicated on the promise of safeguards, and even of benefits, that are

---

140 R. Spinello, 'Deontological Ethics', in *International Encyclopedia of Ethics*, ed. John K. Roth (London: FD, 1995), 219.
141 Dietmar Mieth, 'Why We Cannot Allow the Cloning of Human Beings', *Theologische Quartalschrift* 177 (1997), 1–6.

most likely unattainable. The second is an unconditional rejection, a defensive protective move in the face of abstract endorsement. This could also be viewed as abstract and could be an unstable source of resistance. The third is an equally strict rejection of the technology but in the light of a cluster of concrete relative arguments (philosophical and theological) which may or may not be predicated on categorical arguments. Relative arguments, he says – those that cannot totally dismantle every counter argument – can act to hold a strong moral position, a ban on human cloning in this case. These relative arguments allow for further debate; individual arguments are still open to discussion 'under changed assumptions, contexts, and consequences'.[142]

However, he points out that although this provisionality is self-evident it should not lead to the conclusion that ethical provisions will be revised in future to serve some political or technological expediency. They could however be revised in light of new evidence.[143] A corollary of this, for Mieth, is that not all arguments need to be categorical to be worthwhile, although in the case of human cloning the categorical argument against cloning stands.[144] A cluster of arguments, none of which may be categorical, may offer sufficient justification for an ethical position. This is relevant in the domain of environmental ethics where the dignity of the human person may not be at stake. In that case a string of strong relative arguments, for or against the use of a technology, rather than one categorical argument, can have great normative force.

A commitment to a public shared moral discourse

Like many positions in theological ethics, the autonomy position takes seriously the achievements of philosophy and science without either claiming to hold the trump card or playing down its own insights. It specifically seeks to include the insights of domain-specific contexts and

---

142 Ibid., 6.
143 Mieth, 'Bioethics, Biopolitics, Theology', 10.
144 Ibid.

to examine the links between general moral knowledge and particular moral choices. The autonomy approach is committed to interdisciplinary discourse. For these reasons it can be considered committed to public dialogue and engagement while yet aware of the limit to, and critiques of, instrumental reason and progress in a pluralist world.[145]

The hypothesis that freedom exists, with its emphasis on contingency, on finiteness, has potential in interdisciplinary dialogue. Finitude is considered to be a concept of convergence between theology and philosophy'.[146] It makes us question the possibility of human perfectibility and the ability to solve ever greater problems by escalating technological solutions. It also acknowledges the possibilities that lie beyond a narrow rationalistic philosophy or contextual blindspots in the justification for and legitimation of our conclusions. It explicitly leaves room for such questions as meaning, the value of the human person and the value of our lifeworld.

Mieth also recognises the need for theological insights to be translated or transposed into philosophical argumentation in his conductive method. The biblical understanding of the human person as being made in the image of God can be inspirational in ethics and belongs to the context of Christian ethics, but it is not a substitute for arguments that need to be mediated philosophically.[147] In emphasising this need for translation Mieth is not suggesting that the 'judgement is achieved by philosophical means'. What he stresses is that the significance of biblical narratives in ethics, and of theological insights in philosophy, are not self-evident and need elucidation and clarification. Translating into philosophical categories has the advantage of allowing the discussion get beyond what might be viewed as purely illustrative, in relation to narrative, to questions of criteria and their application.[148] It also makes explicit

---

145 Cf. Johann Baptist Metz, *Faith in History and Society* (New York: Crossroads, 1980).
146 Mieth, 'Bioethics, Biopolitics, Theology', 20.
147 Cf. Ibid.
148 Dietmar Mieth, 'Auf Dem Wege Zur Ethical Correctness: Vorbemerkungen Zur Ethik Des Libensmittelrechts Am Beispiel Der Novel Food-Problematik', *ZLR*

the moral insights that philosophy owes to theology, a project close also to O'Donovan's heart.

Translation and transposition carry, however, a proviso for the Christian autonomist, in that theological concepts cannot be translated without remainder into philosophical currency. As Junker-Kenny points out, understanding ourselves as being made in the image of God may have parallels with the concept of human dignity established on purely philosophical grounds, but it is not exhausted by it. 'The thought that we are made in the image of God should be both more disconcerting and more uplifting that being a religious parallel to insights which can be gained on secular grounds.'[149] Thus, moral arguments cannot replace the theological insights partly expressed in them. Not to 'translate' at all, however, abandons the universalist outreach of the Christian message of salvation and a theological anthropology of freedom receptive to this hope.

In a similar way, the thought that creation is good and destined for salvation, and that 'creatureliness' is an abiding aspect of human anthropology, conveys more than that we should engage in environmental conservation and preservation, or indeed simply celebrate 'embodiment'. The categories 'creation' and 'creatureliness', are difficult to translate in environmental theology. They carry semantic and moral potential that cannot be simply reduced to questions of environmental concern. The givenness in creation captures more than the 'spontaneity' in nature. From a faith perspective, the fact that creation is not of human making but that it nonetheless demands our response is not alienating, or indeed all-consuming. The world is affirmed as the appropriate place for the human person, the graced creature of a benevolent creator. We are not therefore subject to the vagaries of the cosmos, nor are we condemned to live with our 'immortal longings' in an alien habitat of endless cyclical processes. A Christian interpretation of nature as creation guards

---

Zeitschrift für das gesamte Liebensmittelrecht 3 (1999), 278. Translated by Derry Cummins. Further page numbers are in the text.

149 Maureen Junker-Kenny, 'The Image of God – Condition of the Image of the Human', in *The Human Image of God*, ed. H.G. Ziebertz, et al. (Leiden: Brill, 2001), 81.

against any 'triumphalist environmentalism' that seeks to preserve nature, even at the expense of the human person. It also guards against 'despairing' positions that are blind to the creative possibilities that suggest that humanity can flourish, not despite nature, but in its very responsibility for it. In chapter four I will examine how Christian theology interprets nature as creation.

The autonomy position in Christian ethics is committed to contributing to ethical discourse in the public domain and reappropriating 'secular' insights for theological ethics. It has an openness towards the world, which is predicated theologically on the universal offer of God's grace and takes the 'secular milieu' as a 'genuine source of theological knowledge'.[150] It does so, however, not by reducing theological categories to philosophical categories but by continuing to retrieve the ongoing potential of its own sources and value frameworks.

Non-human creation in a Christian reconstruction of autonomy

A Kantian-inspired moral theology, in its Copernican turn, reconstructs our insights into the world critically, from the perspective of the human subject. Since it is anthropocentric, in that sense, it may seem an unlikely ally in the search for an environmental ethic. Yet for Kant nature does inspire awe and an appreciation of beauty that is morally significant for humans. In Mieth's reconstruction of Kant, we find the non-human creation defended both in the concept of indirect duties to nature and in an inclusive rewriting of the categorical imperative. In the autonomy perspective the onus is not on non-human nature, personified in some way, to assert its rights or claim intrinsic value. Rather the onus is on the human person to recognise non-human nature in its continuity and difference and to act in a way towards it that is in keeping with human dignity.

A reading of Kant that was exclusively focussed on his moral philosophy would lead us to assume that nature is merely a backdrop for finite

---

150 De Schrijver, *Recent Theological Debates in Europe*, 121.

freedom. Seeking a moral significance for nature in Kantian philosophy may seem, at its best, to be simply misguided. Any duties we have towards animals, for example – who do not have intrinsic dignity but only conditional worth – are only 'indirect duties towards humanity'; treating animals cruelly is morally significant only because it reflects badly on us as rational agents. Kantian philosophy, from this perspective, seems inadequate from the outset in that it is irredeemably 'anthropocentric'.

However, indirect duties to animals, and by extrapolation to the non-human world, may well constitute an adequate environmental ethic. It gives categorical, unconditional value to the human person while also recognising that we have duties, and that is no small thing, nor to be feared, to the wider reality of which we are part. The natural order is significant because it is the context in which 'rational nature' lives: if we value in ourselves the conditions that make 'rational nature' possible then this has implications for non-human nature.

Onora O'Neill helpfully points out that nature and morality in Kant do not represent independent worlds, and although we cannot integrate them easily, we nevertheless must understand them to be compatible. In *Groundwork,* Kant, she says, indicates that ...

> ... nature and freedom do not belong to two independent worlds, or metaphysical realities, but rather constitute two 'standpoints'. We must see ourselves both as parts of the natural world and as free agents. We cannot without incoherence do without either of these standpoints, although we cannot integrate them, and can do no more than understand *that* they are compatible.[151]

The implication is that the human person 'fits into' the world rather than being incompatible with it. For O'Neill 'a commitment to acting morally in the world depends on assuming (postulating, hoping) that the natural order is not wholly incompatible with moral intentions'.[152]

---

151   Onora O'Neill, *Autonomy and Trust in Bioethics* (Cambridge: Cambridge University Press, 2002), 181.
152   Ibid.

This is consistent with respect for a reality that humans did not create or construct for themselves.

An autonomy framework, emphasising human dignity and freedom, would not capitulate easily to environmentalism. Nor does it need to. It need not be assumed that it is an abstract rational nature rather than an 'embodied' rational nature that is the subject of Kantian philosophy. The unconditional value of the human person is not an obstacle to developing a consistent and sustainable environmental ethic.

In the 'second generation' theological reformulation of the autonomy position – which involves a reintegration of normative ethics and an ethic of the good life – it is possible to readdress the question of the moral significance of nature. Mieth attempts to name some normative ground rules in environmental ethics by broadening the reference of the categorical imperative to include a recognition of the integrity of non-human nature in both the development of human institutions and the recognition of human interdependence in embodiment or corporeality.[153] His aim is to find a formulation of the categorical imperative that integrates 'fragmented perspectives' on human moral autonomy and in particular the natural and environmental aspects of human life.

Mieth offers several environmentally sensitive restatements of the categorical imperative in order to voice some of the 'partial perspectives' of embodied reality that are philosophically neglected. His first formulation is, act 'in such a way that the life necessities of non-human nature be maintained and developed as the place of contact with the contingent corporeality of the human being' (12). In this formulation he stresses the continuity between human corporeality and the natural world. His second formulation is, act 'in such a way that the instrument of a peaceful and creative self-realisation of the human being does not endanger its physical and biological resources, but attempt to relate to the human being in accordance with the inherent characteristics of those resources' (12). Here he stresses the obligation to appropriately maintain the integrity of

---

153 Mieth, 'Auf Dem Wege Zur Ethical Correctness: Vorbemerkungen Zur Ethik Des Libensmittelrechts Am Beispiel Der Novel Food-Problematik', 278.

resources, since it is in the power of the human person, and not of other organisms, to wilfully endanger biological life. Lastly, he states, act 'so that the human institutions serve the development and preservation of personal human corporeality in such a manner that the intrinsic value of the 'prehuman' world is preserved, reconstituted, and promoted to as great a degree as possible and that the specifically human life in creative autonomy is made possible' (12). Here he stresses the need to build just institutions, or to transform unjust institutions, that first serve human life and creativity, and not only conserve, but also allow the non-human world to flourish.

For Mieth the value of the natural world that is appreciated teleologically also belongs, deontologically, to the concept of human dignity itself. Rather than seeking to abolish anthropocentrism, he rewrites the anthropology on which it is based. Human beings have the power to alter and destroy the world of corporeality or to meet their obligation to respect it by respecting it in and around themselves. Mieth does not neglect the theological dimensions that can be credited with this sensitivity for nature outside of human nature. Nature has its own integrity because it is the creation of a good God and is not only at the disposal of humanity. Creation theology emphasises the goodness of the created order and offers the possibility of salvation for all of God's creation. Undoubtedly the creation narratives also show the negative aspects of creatureliness. However, most significantly they put the onus firmly, and anthropocentrically, on the human person to live up to being made in the image of God, even given finitude and fallenness, rather than on the non-human creation to justify itself in the face of environmental injustice which betrays the task of stewardship.

Summary

Divine command ethics gives priority to biblically revealed morality. It is theologically deontological; having a strong theory of obligation based on divine revelation. Its aim, an aim shared in some way with all Christian ethical positions, is to trace moral concepts back to their roots in scripture. However, it understands itself as almost exclusively Christian in its appeal and for that reason is not always easily translated outside of communities of faith. It can carry with it a separatist implication, a refusal to engage in public discourse. It is also prone to the motivational fallacy; assuming that a good motive could substitute for reasoned analysis and argumentation. And it can mask the philosophical assumptions it necessarily employs in ethical discourse. Although, as O'Donovan points out for Christian ethics, the public realm also fails to recognise or acknowledge the debt that modern society owes to Christian theology and ethics.

As far as Christian ethics is concerned, the reliance on Scripture as the axis of revelation gives it an air of security which cannot always be claimed by other positions. With the emphasis on Scripture, divine command has the advantage in that its primary resource, the Bible, brings to our attention the glories of the natural world, the wisdom and balance in nature and the place of humanity in the created reality: not as the centre, but the most communicative and responsible point of contact with the Creator. Having said that, however, it is not an obvious ally in the search for an environmental ethics that is sensitive to the natural world, since it subordinates creation to redemption and fails to work out the link between revelation and creation.

However, in the ethics of Oliver O'Donovan we see a rapprochement of a 'natural' and an 'evangelical' ethic that has positive connotations and resources for a Christian ethic of nature as creation. Like Mieth he sees the Gospel as more than an ethical textbook, it is rather a specifically theological and religious answer to the question of meaning. Biblical theology requires a hermeneutical sensitivity and a respect for the historical-critical approach but, as in O'Donovan's work, it also requires a theological imagination that does full justice to a view of reality as we encounter it, as the creation of a good God. O'Donovan's ethic does not rule out a methodological separation between questions of morality and questions of faith and indeed recognises the autonomy of the 'order of creation' as a gift already given. O'Donovan's theology rebuts the implication, common to much philosophical ethics and to ethics in the public realm, that religious commitment is esoteric, inaccessible, inexplicable and simply heteronomy.

His rapprochement of a realist principle and an evangelical ethic, opens up the possibility of an environmental ethics that does not play down the significance of the non-human creation in the story of salvation. However, if he has made room for the non-human creation in salvation he has damned human creative potential perhaps further. He draws a sharp distinction between the natural and the artificial, between the goodness and spontaneity of non-human nature (viewed as creation) which he rightly emphasises, and the artificial and less loveable world of human making. As a critique of instrumental reason his analysis stands. However, as we will see in chapter four, this comes to be interpreted, in environmental ethics, as a plea for an almost static 'preservation' of non-human nature in its 'original created order', as we find in the work of Michael Northcott. This downplays the role of all human activity and neglects the ongoing creative significance of God's relationship with the human and non-human world.

Communitarianism offers a reinstatement of the value of community in moral discourse after a long sojourn in the wilderness of atomistic individualism. It argues for the role of community in moral formation and ethics. In its ecclesial form it is in danger of conflating all possibilities for morality with Christian ethics, making unnecessary adversaries in moral philosophy and in the public realm. It could be defended for maintaining its prophetic role as a beacon in an otherwise dark age. Yet that is to underplay its potential dangers and at the same time, as Porter rightly points out, to patronise it.[154] Retreating into an exclusively church

---

154 Jean Porter, 'Perennial and Timely Virtues: Practical Wisdom, Courage and Temperence', *Concilium: Changing Values and Virtues* 3 (1987).

setting, and turning a back to the world is part of the danger. Solving the questions of the apparent loss of meaning in modern culture by demonstrating the functional value of religion and, at the same time, silencing the churches on public debate seems unworthy to the theological task of Christian ethics.[155] Religion, from the ecclesial communitarian perspective, serves to compensate for societal dysfunction while at the same time remaining aloof to criticism from without. Its attitude to public debate in general and to moral philosophy in particular is at least ambiguous if not hostile. Any shared, community morality is by and for the community which shares it.

A critique of atomism, however, need not mean a rolling back of the Enlightenment. Communitarianism is a corrective in that it envisages ethics as a commitment to a particular form of life in community, placing the emphasis on socialisation into a community's conventions. This has several advantages, not least that of being inclusive, accepting all members equally and indeed making the most of any shortcomings.[156] Communitarian approaches have also prompted a reconstruction of Aristotelian virtue ethics and examined the role of virtues in the formation of moral communities. Moral virtues, communitarians contend, are required to live a moral life and the 'virtue' of practical reason involves an 'adequate sense of the tradition to which one belongs'.[157] Although virtue ethics may not be comprehensive as an alternative to other ethical theories, it has broken the grip of a traditional choice between deontology and utilitarianism in ethics.[158] The communitarian insistence on the rehabilitation of virtue ethics has stimulated a positive discussion of the role of community in moral formation and biographic identity, across the confessional divide. We shall see how Celia Deane-Drummond, in chapter four, develops a virtue approach for an ethics of nature.

---

155 Fergusson, *Community, Liberalism and Christian Ethics*, 154 & 69.
156 Stanley Hauerwas, *Suffering Presence* (Edinburgh: T. & T. Clark, 1988).
157 Kerr, 'Moral Theology after Macintyre: Modern Ethics, Tragedy and Thomism', 35.
158 Ibid., 33–34.

Natural law is of particular significance in its validation of the creation and human reason; in its development of a social and political thought that thematises entities intermediate between the state and the individual; and in its realist ethic that allows that nature is in some way morally relevant. The social concepts of the common good and of subsidiarity already address the central elements of communitarian critiques of atomistic individualism. The concept of the common good registers the intersubjective, social aspects of human anthropology, which are constitutive of, rather than external to, the human person. This goes far beyond simplistic notions of average utility or preference satisfaction. It also captures more than the notion of the public good, which itself is only one aspect, if a crucial one, of the common good. The principle of subsidiarity has significant potential for the realisation of environmental goals in ethics since, as a mediating principle, it helps to identify the most local level at which an ethical question needs to be addressed.

The autonomy position however gives the clearest account of the interaction between the different constituents in ethical discourse. It allows a distinction to be made in Christian ethics between the sources and content of morality, it demonstrates the need to make our philosophical and theological resources and themes explicit. The Christian autonomist position, as exemplified by Mieth, goes beyond an appropriation of Kant to consider not just the normative questions in morality (the relationships between people's actions and institutions) but also to consider virtue ethics, the ethics of good counsel, of narrative and biographical identity and that which contributes to human flourishing. It retains however its deontological imperative and is modern in respect of acknowledging the conflicts between (and within) humans that lead to domination. Against this human temptation it recognises the other as an equally original freedom.

This developed autonomy position also makes clear the links between deontology and teleology in moral discourse. Divine command morality, natural law and autonomy all work as deontologies; divine command deontology is revelation based; the deontological aspect of natural law is based on a metaphysical assumption about human nature with an undeclared but in-built theological assumption. In contrast the autonomy position – based on a formal analysis of reason and the dignity of the human person predicated on the capacity to be moral – retains a 'metaphysics of morals'; a metaphysics that postulate the existence of God, at the limits of reason. In this way the autonomy position in theology can methodologically separate the question of the source of obligation in morality and the question of faith. For a Christian autonomist, the teleological aspects of ethics need a deontological framework so that Christian visions of the good life are not themselves instrumentalised and sacrificed to prevailing prejudices.

In Christian ethics the Gospel is morally relevant and normative but it is concerned with more than simply ethics.[159] It addresses not alone questions about how to live a morally just life but also the question of meaning. Divine command could be accused of reducing all of religion to ethics in the insistence that one must spring only from the other. It also seems to impose or to discover a unity in what is really a diverse body of literature, when referring to Scripture.[160] Divine command morality highlights the limits of reason and the danger that the gospel may be used merely to confirm prevailing moral positions. However, it has no refuge from a dependence on reason other than a retreat into authority or into eschatological criticism.[161] If reason cannot arbitrate then we are left to either accept the authority of Scripture or the teaching authority of the church or hope that it all works out in the end.

The corollary of this identification of Christianity with ethics, and ethics with the Word of God as revealed in Scripture, is the identification of all of Christianity with the Bible.[162] Yet the Bible is not the sum total of Christianity or of ethics. Neither are divine command moralists insulated from having to interpret in order to apply biblical ethical injunctions, as O'Donovan shows. They also have to construct their personal canon within the canon. However, the prior bias in favour of a specifically

---

159   Mieth, 'Autonomy of Ethics – Neutrality of the Gospel?'.
160   Sean Freyne, 'The Bible and Theology. An Unresolved Tension', *Concilium: Unanswered Questions* 1, no. 1999/1 (1999), 19.
161   Mieth, 'Autonomy of Ethics – Neutrality of the Gospel?', 34.
162   MacNamara, 'The Distinctiveness of Christian Morality', 104.

Word-based ethic views philosophical mediation as disloyalty. In contrast, rather than being disloyal, from the autonomy perspective this is a necessary task in theological ethics.

An environmental ethic or an ethic of production from an autonomy perspective would need to integrate several aspects. At the deontological level Mieth develops an inclusive version of the categorical imperative. Rather than assuming that anthropocentrism is contrary to an environmental ethic, the autonomy position does not let the human person off the hook too easily. It is through a revised anthropology–one that registers human corporeality and obligation–that 'liberation' for non-human creation can be envisaged. An ethics of obligation also needs reinforcement at the level of values, so that our obligations can be realised. The realisation of a goal, such as sustainability in environmental ethics, requires both an overarching framework of normative principles as well as contextualisation at the level of institutions and social structures. The Christian context is not external to these tasks. It provides heuristic, sensitising and motivational potential and, as a hermeneutical tradition, the tools to integrate and relativise ethical insights across several disciplines.

To develop an environmental ethic, or an ethic of production, we next have to examine how different theologies of the environment have emerged in response to the so-called environmental 'crisis'. We also need to refer to the related question of poverty, an issue that environmental ethics, with its emphasis on non-human nature, has a tendency to mask. Theological ethics must examine the validity of the claim that there is a conflict between sustainability and development in the developing world and between sustainability and permanent economic growth in the developed world. Only then can it hope to develop an ethics of production that respects the dignity of each human person in their corporeality and to find just mechanisms that help to institutionalise the conservative use of non-renewable resources.

CHAPTER 3

Environmental Theologies and Reconstructions of Stewardship

Christian theology has responded to the contemporary concern for 'nature' in light of the evidence that humans in modernity have an impact on their environment that is unprecedented, destructive and frequently irreversible. Environmental ethics is practical in its orientation but it has a background in systematic theology, in both cosmology and creation theology and in philosophical and theological anthropology. Chapter four will focus on some of the systematic aspects of creation theology that provide a theologically informed reconstruction of creaturely relationships, in light of contemporary environmental questions. Here, in contrast, I will concentrate on some specifically heuristic and practically oriented models in environmental ethics and in biblical theology.

Environmental consciousness has prompted theologians to restate and re-emphasise two central insights of creation theology – the goodness of creation, and the radical distinction between the Creator and the creation. The first, the goodness of creation, was defended from the early centuries against the dualism of Gnosticism, the second against the monism of pantheism. Gnosticism disparaged the material creation and implied a dualism in the Godhead – between true divinity and the creator of the material world, the demiurge. This was antithetical to the Christian inheritance of Jewish monotheism and its affirmation of the created order. Monotheism emphasised the radical distinction of God from God's creation, against the monistic inclinations of pantheism. It thus demythologised the cosmos, in the interests in particular of divine and of human moral freedom.

Having laid out the systematic boundaries of Christian creation theology, it is, however, no straightforward matter to read directly from the insights of creation theology to nature as it is understood in the natural sciences. Nevertheless, both secular and Christian calls to replace excessively anthropocentric attitudes to the earth have led to two responses in 'environmental' ethics. The first response is to explore the potential of non-hierarchical relationships, between the human and non-human natural world, and is termed biocentrism or ecocentrism.[1] The second response, theocentrism, argues for the necessity of a renewal of all creation's relationship with God, as the only way in which non-human nature can be defended against human instrumentalism.

Ecocentrism defends a non-instrumental value for the non-human natural world. However, where it advocates a simplistic holistic synthesis with nature it does not easily sit with ethical monotheism as understood in the Judeo-Christian vision. Christian creation theology is intrinsically anthropocentric, albeit in a specifically Christian way. Nevertheless, where ecocentrism is understood as a 'consciousness movement', as Palmer suggests – rather than as a form of misanthropy – it is a welcome corrective to instrumentalist interpretations of anthropocentrism.[2]

Theocentrism, the second strand in theological environmental ethics, reasserts the need for a 'religious' attitude to the cosmos as the creation of a good God. Proponents suggest that a realignment of thinking 'back' to God will deliver an environmental dividend by overcoming the (false) dichotomy between anthropocentrism and ecocentrism. The theocentric model calls Christians to turn away from the traditional anthropocentrism of Christianity, which is viewed as overly instrumentalist, to a creation-centred ethic.[3] This is a theme in many Christian responses to the 'environmental crisis'.

The need for a theological foundation, or at least some kind of metaphysical foundation, for an environmental ethic can be found not only in Christian theology but also in the 'secular' positions of deep ecology. Deep

---

1 Biocentrism refers to living organisms, while ecocentrism refers to the totality of the natural world.
2 Clare Palmer, 'A Bibliographical Essay on Environmental Ethics', *Studies in Christian Ethics* 7, no. 1 (1994), 91.
3 Porter, *The Recovery of Virtue*, 24.

ecologists claim that a concern for the material or natural world needs to be underpinned with a metaphysic.[4] In some cases, undoubtedly, such attempts seem to amount to a naïve instrumentalising of religion and have little to do with Christian theology. Yet, it may come as no surprise that we find a theologian-scientist like Celia Deane-Drummond suggesting, in the context of the 'new genetics', the importance of a theological interpretation. In relation to transgenic food in particular, she suggests that theological perspectives 'may now be indispensable in helping explain to largely secular institutions the sources and dynamics of conflicts now threatening to paralyze the development of what is being posited as a key technology for the twenty-first century'.[5] Theocentrism seems to be able to explain those attachments and beliefs that resist a technical attitude.

Aside from analysis of ethical models, environmental theologies also need to develop insights from New Testament sources. I will open this chapter by presenting some insights from two aspects of NT biblical theology that can contribute to the development of a Christian environmental ethic; the salvation of all creation in Romans 8. 19, and a new reading of Jesus' ministry, which reconstructs Jesus' self-interpretation of his own scriptural heritage. The work of Sean Freyne is presented here in some detail since Freyne takes up the question of Jesus' interpretation of his own creation scriptures in light specifically of the natural and farming environment encountered in his ministry.

I will defend a Christian environmental ethic that remains firmly anthropocentric but which can take on board the critiques and insights of both ecocentrism and theocentrism. I shall then examine how the biblical injunction to dominion has been reconstructed in theological environmental ethics as 'stewardship'. These reconstructions are worthy resources because they resist dehumanising and reductive approaches to non-human reality and defend cultural achievements that are being undermined by a banal ideology of innovation and progress. I then employ

---

4 Cf. Peter Forrest, 'How Kenotic Process Theology Underpins Humanist Deep Ecology', *The Australasian Journal of Process Thought* 1, no. 1 (2000).
5 Celia Deane-Drummond and Bronislaw Szerszynski, eds, *Re-Ordering Nature; Theology, Society and the New Genetics* (Edinburgh: T. & T. Clark, 2003), 22.

these resources to critique recent proposals to introduce GM crops to Irish agriculture.

Biblical theology and the non-human creation

A systematic presentation of the origins, preservation and the salvation of specifically the non-human creation will be the theme of chapter four. In contrast in this section I hope to outline some of the perspectives on these to be found in recent readings of the NT text. I will first consider briefly interpretations of Paul's revelation of the salvation of all creation in Romans 8: 19. In this passage, Paul affirms the necessary link between creation and salvation in Jesus' teaching. I will then examine Freyne's new reading of Jesus' own self-understanding of the natural environment, in the context of farming and fishing in first century Galilee, which enlightens theological views on the liberation of creation in history. The intention in this rereading is not to project present day concerns onto an ancient text. As Freyne warns, theologians must aim to avoid the pitfalls of past mistakes and the danger of wild misinterpretations of the 'historical' Jesus. My intention is not to just add a 'green' Jesus' to the mix. This rereading is undertaken in the spirit of a 'lively re-assessment of the christological tradition'[6] rather than from any desire to provide 'proof texts' in the interests of an environmental ideology. It represents the beginnings of a shift in readings of the biblical text that interprets nature as creation and as a true context of human life.

---

6   Cf. Andrés Torres Queiruga et al., 'What Is at Stake in Christology', *Concilium: Jesus as Christ* 3 (2008).

Biblical models in theology

Recent biblical scholarship which focuses on 're-imaging nature' in the biblical text has been reviewed by Hiebert.[7] He suggests that this scholarship, which is motivated by a concern for the natural world, is interested in more than just mining the text for its potential to support an 'environmentally friendly' agenda. '[M]ore is involved than the mere retrieval of biblical passages on nature or the affirmation of God as creator and the human as caretaker.'[8] Traditional assumptions are re-examined in the interests of 'remaking' the moral imagination. This is exemplified by a significant shift, from a narrow understanding of nature as non-human and antithetical to human existence, to nature understood as a true context of human life.

Biblical studies reflect the classical distinction between nature religions and the demythologised world of Israel's monotheism.[9] In emphasising the radical distinction between God and God's creation, and between human morality and the cyclical draw of natural processes, it is clear that Israel's religion could not be construed as a nature-religion. God acted through singular events in history and not in the unchanging pattern of endlessly repeating seasons, cycles with no apparent purpose or goal. The ethical element in Israel's religion emphasised liberation and freedom from, and not endless service to, unrelenting matter. In this reading 'history replaces the cosmic' in biblical religion. Israel's religion is read as uniquely historical in orientation and human society is set off from its natural environment. This understanding had repercussions for environmental theologies, not least the tendency to treat nature as a backdrop to human history rather than a true context of human life.[10] 'The principle of liberation has been closely associated with the dynamics of history rather than with the rhythms of nature, and the roots of this liberative

---

7   Theodore Hiebert, 'Re-Imaging Nature', *Interpretation; A Journal of Bible and Theology* 50 (1996), 36–45.
8   Ibid., 44.
9   Ibid., 37.
10 Ibid., 41.

historical consciousness have been traced back to the biblical belief in God's involvement in history above all in the exodus from Egypt'.[11]

The emphasis on the history of salvation may well have relegated non-human nature to the background. However, concern for the non-human creation, as the context for finite human freedom, is a theme that environmental theology, in a changing historical context, wants to bring back into the foreground. It is hoped to get beyond interpretations of dominion as domination or as conformity with a prior order. To that end I will examine two themes that could enlighten a biblical environmental theology namely Paul's Rm. 8: 19 and Freyne's new reading of Jesus' ministry. Notwithstanding the difficulties of making 'normative' use of the biblical text, these, it is hoped, will be instructive for a reconstruction of Christian 'stewardship'.

Romans 8:19 and the 'salvation of all creation'

The significance of Paul's Rm. 8: 19[12] for environmental theology lies in its being the clearest statement in the NT of the possibility for salvation for all creation, including the non-human creation. It is not however a standalone text and it is not isolated in its affirmation. For many biblical scholars it is simply the most lucid statement of the underlying and necessary connection between creation and salvation to be found in the canon. In a yet unpublished paper, David Hutchinson Edgar explores Rm. 8: 19–23 and the unbreakable link between creation and 'eschatology' in the Pauline

---

11 Ibid., 40.
12 Rm. 8. 18–23; 'I consider that the sufferings of the present time are not worth comparing with the glory about to be revealed to us, for the creation awaits with eager longing for the revealing of the children of God; for the creation was subjected to futility, not of its own will but by the will of the one who subjected it, in hope that the creation itself will be set free from its bondage to decay and will obtain the freedom of the glory of the children of God, We know the whole creation has been groaning in labour pains until now; and not only the creation, but we ourselves who have the first fruits of the Spirit, groan inwardly while we wait for adoption, the redemption of our bodies.' (NRSV)

corpus.[13] He defines eschatology as the unfolding consequences or completion of a process that has already begun in the world. Central to this analysis, for Hutchinson Edgar, is the idea that time be understood as an 'extended present' rather that having a strong future orientation; although traditionally Pauline Christianity has been read as having a strong future orientation to the coming of the kingdom of God.

> Paul's continual recourse to the creation-fall narratives in Genesis, as well as to prophetic images of the restoration of harmony to creation, ... is firmly rooted in the past as a source of understanding and experiencing the unexpected ongoing present. (3)

Hutchinson Edgar argues that Paul's letter to the Romans is stamped with intertextual images of origins, in particular with the creation theology of Genesis and of Isaiah. These intertextual references to Jewish scripture are not irrelevant to Paul's task. Significantly, in Romans 8: 18–23 they allow a confirmation of the promised restoration which is not just for all Israel, or all humankind, but for all of creation (2).

The image of the subjection of creation in the Romans passage begs the question; who is the oppressor referred to in the text, 'the one who subjected creation to 'futility'? Is it God, the devil, or humanity? Traditional theology would not indict God, and although the devil makes a convenient scapegoat he is still a part of the created order. Rather it is humanity that is in the dock. As Hutchinson Edgar suggests, the echo in the Romans text is to the creation–fall narrative and to the authority given to Adam to have dominion over the earth. It is mankind that has, in part at least, subjected the creation to its 'bondage to decay'.

Christian soteriology is human-centred but the Romans text presupposes a complex anthropology that is inclusive of the non-human creation. It implies that non-human creation is destined for salvation and liberation from enslavement, alongside, not apart from, humanity. The assumption in the text is that the human person contributes to the

---

13   D. Hutchinson Edgar, 'Back to the Beginning: Paul's Eschatology in Romans', *Trinity College Dublin (unpublished ms)* (2002).

enslavement of creation. If that is the case then we need to reread the text with the salvation and liberation of all creation in mind.

A new reading of Jesus' ministry for the liberation of all creation

Any new reading of Jesus' ministry with the non-human creation in view has to avoid what Sean Freyne calls a superficial and uncritical adoption of current environmental concerns. Yet it also has to make some redress for the subordination of creation to redemption in some Christian theologies in light of Jesus' emphasis on his own scriptural tradition. Creation theology, in Jesus' self-understanding, is foundational to biblical thought and is not 'chronologically late or theologically secondary' as it has sometimes been interpreted.[14] Freyne suggests a revised context with which to interpret Jesus' attitudes and values as regards the landscape and places of his ministry. He develops this by combining textual, intertextual, contextual, archaeological and landscape studies.

Freyne observes that Jesus has so often been interpreted according to the needs of the day; Jesus as moral teacher has been successively supplanted by Jesus the Cynic, by Jesus the magician, by Jesus the wandering charismatic, or Jesus the social revolutionary. In his book *Jesus, a Jewish Galilean* Freyne wants to avoid throwing a 'green Jesus' mercilessly into the fray.[15] Instead he painstakingly contextualises the role that place and landscape played in Jesus' ministry in an investigation of Jesus' social concerns. He examines the complex, so-often inescapable relationships, between natural resources and social structures, from wilderness and desert to fertile valley and mysterious uplands. Freyne seeks to elaborate

14 Hiebert, 'Re-Imaging Nature', 41.
15 Cf. Sean Freyne, *Jesus, a Jewish Galilean: A New Reading of the Jesus Story* (Edinburgh: T. & T. Clark, 2004), 24–59. Methodologically, Freyne wants to avoid erroneous readings of the historical Jesus; as non-Jewish by virtue of his being Galilean – popular in the early twentieth century; as a mere construct of the evangelists; or psychologically in the vein of the what-would-Jesus-do-if-he-were-here-now mode; or indeed in the vein of popular fiction.

on Jesus' interpretation of the places he travelled to and how he read his own scriptures in light of that experience. He finds that Jesus emphasised his faith in the creator God and the traditional scriptures of Genesis, Isaiah and Daniel over Joshua, Ezekiel, Maccabees, Exodus or Judges:

> The argument in this study is that Jesus' primary motivation was based on his understanding of Israel's God as the creator God, and that this perspective determines the manner in which he appropriates certain aspects of his tradition, in contrast to other strands of Jewish thinking in his day.[16]

In his rereading Freyne contends that as someone conversant with his own scriptures, Jesus was neither a marginalised Jew nor a hellenised or romanised Mediterranean, as he has sometimes been read. In relation to confronting the challenges of empire, for example, Freyne reads Jesus interpreting Genesis in preference to Exodus. 'If the suggestion is correct [that Jesus did not call for the removal, conversion or annihilation of occupying forces implying] that his image of God was inspired more by the Genesis than the Exodus account, then we must ascribe to him a broader horizon for understanding Israel's role among the nations than one of hostility, leading to aggressive militancy in reclaiming Israel's political freedom' (135). For Freyne, salvation for Jesus seems clearly not to have been that of the triumphalism of imperial power, or indeed its converse, the revolutionary expectation of imminent and radical change (127).

Perhaps, Freyne suggests, Jesus understood salvation as the 'deferred eschatology' of the sage, who lives by the apparent order in the universe while knowing it to be good (if not itself God). Yet, although Jesus' understanding of the Kingdom was not imperial or revolutionary, neither was it sapiential. Rather, Freyne contends, it was apocalyptic. 'The symbol of the kingdom of God as employed by Jesus signified God's victory over the evil forces, a conception that was shared with standard Jewish apocalyptic expectations. The poor and the oppressed would be vindicated, the wicked and unjust punished' (137). Indeed for Freyne, Jesus was hardly received as a sage as it was traditionally understood. 'The people

---

16   Ibid., 138. Further page numbers are in the text.

of Nazareth were unimpressed with his wisdom, we are told, describing him as a "craftsman" (*tekton*), and thereby disqualifying him as a source of wisdom according to the system of Jerusalem and its scribes (Mk: 6.2–4; Sir.38: 24–39).' (48)

An apocalyptic world view seems, at first glance, to have little to say about the liberation of the non-human creation in the present. If the kingdom of God is imminent, then the poor and oppressed are soon to be vindicated and the wicked punished. However, as Hutchinson Edgar has already indicated, this apocalyptic world view must be seen in light of the ancient near-Eastern understanding of the 'extended present'. So we still need to investigate the extent to which Jesus is indebted to those aspects of his creation theology tradition, and what this implies for the 'extended present'.

Freyne suggests that in order to uncover Jesus' reaction to his environment we first need to look at the inherited traditions of Israel in relation to the appreciation of the gifts of the earth (27). Significantly, the tradition that Jesus aligned himself with was one which gave the assurance of a new creation and salvation for all. Freyne sees that

> ... the assurance of the new creation of which Isaiah in particular speaks (Isa. 65:17, 66:22) and which finds its way into the New Testament through Paul (Rm. 8:18–25), indicated that in the Biblical perspective human redemption can only be considered in conjunction with the redemption of the earth itself ... as expressions of God's creative goodness. (34)

For Jesus the salvation promised by the God of Genesis and Isaiah was destined for all creation.

In his ministry Jesus travelled from his desert beginnings with the ascetic Baptist to the fertile lower valley and then on to the uplands of Galilee, the mysterious wellspring of this fertility. Although he began with John in the desert, Jesus' ministry took on a different style in Galilee in response, Freyne suggests, to a changing social circumstance which he experienced there. From Josephus we get a picture of Galilee as an effortlessly bountiful land:

> For the land is everywhere so rich in soil and pasturage and produces such a variety of trees that even the most indolent are tempted by these facilities to engage in agriculture. In fact it has all been cultivated by the inhabitants and there is not a single portion left waste. (39)

This bountiful picture, Freyne indicates, is borne out by archaeological, geographical and toponymic evidence.

Yet we know that Jesus' declarations of beatitude were addressed to people who in fact faced material impoverishment and who saw the promise made to their ancestors as being enjoyed by others. How could this situation in life be interpreted as a blessing? The contrasts that Jesus experienced as he travelled from the bare harshness of the desert to the bountiful valley, had the potential to be a reminder of the blessedness of life, and this is what he preached. Jesus interpreted this blessedness as a graced moment, to be savoured, rather than one of awaiting judgement (42). To do this, Freyne suggests, demanded an imaginative development in Jesus' reading of his own scriptures; 'To name this situation as a blessing, and not as a curse, as the Deuteronomic theology would have suggested, called for a bold religious imagination' (47). Jesus' sayings and his itinerant lifestyle point to a total confidence in God's care, one that was not always welcomed by those he preached his message to. In Caphernaum he challenges the values of both farmers and fishermen, fishermen who indeed with their salting industry were adding value to a resource that they retrieved from a bountiful 'nature' as hunter-gatherers rather than as sowers and reapers (50).

Jesus, however, does not even read the mysterious caves and uplands, which may have represented the source of water and therefore fertility for the valley below, as themselves divine. Rather they had come from the hand of God, the God who declared that what he had made was good. He saw them more as an expression or sacrament of God's own goodness. For Freyne, Jesus aligned himself with a demythologising of the natural world but not the complete desacralising of it (37).

The ethical implications of Freyne's reading for the liberation of the non-human creation are not new in one sense, but newly confirmed. From the perspective of Jesus' creation-centred beliefs, the fate of humans and the earth were inextricably bound together. Jesus was not a stoic pantheist, which in the end was a philosophical stance which only supported the prevailing hegemony. He rather represented the prophetic tradition of Isaiah (drawing on Genesis), that challenged the legitimacy of rulers who plundered both human and non-human resources at will.

> The parables of Jesus are such successful religious metaphors because they are products of a religious imagination that is deeply grounded in the world of nature and the human struggle with it, and at the same time deeply rooted in the traditions of Israel which speak of God as creator of heaven and earth and all that is in them. (59)

Summary

Freyne captures the beginnings of a shift in readings of the biblical text – prompted by contemporary environmental concerns – from a narrow understanding of nature as non-human and antithetical to human existence, to nature understood as a true context of human life. Intertextual evidence in the reading of Romans tells us that the goodness of creation is a gift that can be respected or violated by the human person, but which is destined for salvation. Implicit in that text is a complex anthropology that understands that the liberation and salvation of non-human creation is intimately connected to that of human beings. Freyne shows us that the recognition of the giftedness of creation, human and non-human, is not a contemporary projection onto the biblical text but is already explicit in Jesus' self-understanding of his own scriptural tradition. Jesus interpreted the bounty he saw, as he moved from his ascetic sojourn in the desert to the fertile valleys with their farmers and fishermen, as a gift, even in the face of hardships and poverty.

Models for environmental ethics

Theological responses to environmental problems have generally fallen into three broad typologies, anthropocentric, ecocentric and theocentric.[17] Both anthropocentric and ecocentric views of nature have philosophical counterparts. In theology, however, the focus is somewhat different and these typologies represent angles from which to view not nature as such, but createdness. Anthropocentrism views creation from the perspective of humanity, from its position as steward of creation. It is the prevailing paradigm in both philosophical and theological environmental ethics and the prevailing language of global environmental treaties. Ecocentrism, in contrast represents the perspective of the non-human creation, which is often personified, if not always intentionally. Finally, theocentrism argues from the perspective of faith in the creator-God who cares for all creation; a perspective, it could be assumed, that would overcome the negative aspects of all other partial perspectives.

Environmentalists have taken Christianity to task for its anthropocentric biblical sanction of human 'dominion' over the non-human creation. It is argued that this injunction is behind the wanton human destruction of the global environment.[18] Although environmental theologies have taken this criticism on board, it cannot be assumed that the anthropocentrism implicit in Christian theology is automatically at the root of the 'ecological crisis'. Anthropocentrism can encompass a non-instrumentalist approach to nature without personification and without downgrading the value of the human person, precisely by drawing attention to nature's separate status.

---

17 M. Northcott refers to these three as humanocentrism, ecocentrism and theocentrism. Cf. Michael S. Northcott, *The Environment and Christian Ethics* (Cambridge: Cambridge University Press, 1996).
18 The classical statement of this case is taken to be White's 1967 essay on our ecological problems. Cf. White, Lynn, 'The Historical Roots of Our Ecologic Crisis', *Science* 1967: 155 pp. 1203–1207.

Even given the criticisms levelled at anthropocentrism, Christian theology cannot uncritically embrace ecocentrism. The emphasis in theology on the dignity of the human person implies that theology would resist 'deep ecology' where it is found to be contemptuous of humanity. Ecology and ecocentrism have their own drawbacks, not least their tendency towards misanthropy. One early supporter, for example, commented that 'the more misanthropy there is in an ethical system, the more ecological it is' and went on to calculate that, in the scheme of things, the human population in total should be no more than about twice that of bears.[19]

To avoid the drawbacks of both anthropocentrism and ecocentrism, environmental theologies have largely adopted a theocentric approach. In this section I will outline the scope, limits and implications of taking an anthropocentric or an ecocentric approach, and add insights from feminist theology and body theology. I will also assess the potential of the theocentric approach for the development of an environmental theology that is compatible with the autonomy perspective in Christian ethics. First, however I turn to some issues of nomenclature in environmental ethics and theology that need some clarification.

Concepts and terms

It is true to say that if ordinary, everyday history is the sphere of God's action then theologians cannot be silent on non-theological issues.[20] It is not, however, immediately clear how theological ethics should align itself with positions in contemporary environmental ethics; either in its analysis of the so-called 'environmental crisis' or in adopting 'ecological' principles as the ethical answer to environmental questions. Certainly there is no doubt among theologians of the profound, irreversible and negative effect human activity has had, and is having, on the natural environment. Yet, there is less theological evaluation of the normative

---

19 Palmer, 'A Bibliographical Essay on Environmental Ethics', 84.
20 De Tavernier, 'Human or "Secular" History as a Medium for the History of Salvation or Its Opposite: "Outside the World There Is No Salvation"'.

status of this 'crisis'. It is also often taken for granted in environmental theology, that human attitudes to nature can automatically be indicted as instrumentalist, simply because they are anthropocentric. Even to use the term 'environment' to refer to non-human reality, is already to patronise it, as Oliver O'Donovan suggested.[21] Theologians have, in light of that, preferred to use 'ecology' and 'ecotheology' rather than the 'environment'. In this section I would like to evaluate this preference for 'crisis' and 'ecology' in theological ethics.

The environmental 'crisis'

The negative environmental impact of humans on the non-human world is often generically referred to as 'the environmental crisis'. For the theologian Michael Northcott 'the existence of an ecological crisis is increasingly recognised as one of the defining features of life in the late modern era'.[22] Like many authors in this area, he enumerates and quantifies the problems and causes: the mass extinction of species: global warming through the emission of greenhouse gases: the depletion of the ozone layer in the stratosphere: the negative effect of pollutants from domestic waste: industry and farming; and the loss of habitats through soil erosion and desertification.[23]

The expression 'environmental crisis', however, has become a label for all environmental issues; from desertification to the loss of wildflower meadows and urban decay. Yet these are not all commensurate in their impact or in their scientific or political resolution. The depletion of stratospheric ozone through the complicated chemistry of ozone destruction may be, in practice, easier to abate, delay or reverse than global warming through emissions of greenhouse gases. They are both globally relevant and have serious implications, but there may be a greater degree of flex-

---

21 Cf. O'Donovan's comment cf. page 85.
22 Northcott, 'Ecology and Christian Ethics', 209.
23 Ibid.

ibility in the systems or institutions that influence one and not the other.[24] They are related in a non-specific way but attempts to deal with these two questions require different enterprises indeed.

So, although at a general level it is possible to recognise the interrelatedness in all physical processes and in that way to speak of an environmental crisis, distilling every human use of the environment into a 'crisis' fails to make worthwhile distinctions. This is not to deny that grave problems exist nor that decisions made now will not leave a significant legacy or indeed that actions taken (or not taken) will be irreversible. It is clear that resolution is not always possible; species extinction for example is not 'verifiable by repetition'. It is a shocking reminder of our contingent knowledge and power. However, the danger with naming a crisis is that a 'normativity' is given to 'crisis' and there is a foreclosure on other ways, perhaps more productive, of conceptualising environmental questions.

Undoubtedly theologians, in naming this a 'crisis', often want to refer to more than problems associated strictly with the extra-human environment.[25] Michael Northcott argues, for example, that

> ... the environmental crisis is intricately connected with the human crisis of moral ecology and of unemployment and ... the demise of economic justice and moral purposiveness in the global order is directly related to the increasing pace of environmental breakdown.[26]

Whether such moral purposiveness or economic justice ever actually existed is open to question. Nonetheless, he is not alone in the seemingly commonplace realisation that the welfare of humanity is somehow tied

---

24   Cf. Scott Barrett, 'Montreal Versus Kyoto: International Cooperation and the Global Environment', in *Global Public Goods; International Cooperation in the 21st Century*, ed. Inge Kaul, Isabelle Grunberg, and Marc A. Stern (Oxford: Oxford University Press, 1999).
25   Cf. Sean McDonagh, *Passion for the Earth: The Christian Vocation to Promote Justice, Peace and the Integrity of Creation* (London: Geoffrey Chapman, 1994). and Celia Deane-Drummond, 'Development and Environment: In Dialogue with Liberation Theology', *New Blackfriars* 78, no. 916 (1997).
26   Northcott, *The Environment and Christian Ethics*, Preface, xiiiff.

up with the welfare of the extra-human world and that the welfare of the extra-human world has repercussions for the human life world. However, from this social-environmental perspective, environmental problems are multiplied and made even more complex. Northcott cites widely disparate factors as causes, such as overpopulation,[27] limits to economic growth, the pursuit of the ideology of progress, scientific method, Christian doctrine and Cartesian dualism. For Northcott, all of these problems stem from one original cause, namely 'the disjunction between rationality and embodiment; nature and culture which is said to have originated in the philosophy of the enlightenment'.[28] The environmental crisis, in his reading, codes for all the social ills that stem from human fallenness.

However, Christian responses to environmental issues are not univocal; there is more than one theological way to conceptualise the origin and meaning of, or the potential for dealing with environmental questions. It is unhelpful to name all of these issues as an 'environmental' crisis, just as it is unclear that dualisms between nature and culture, between embodiment and rationality, are the simple root of all social ills. The task for modern theology is to positively appropriate those theological and secular intellectual resources that are in keeping with Christian ethical principles, rather than characterise them as fundamentally deficient.[29] To that end it looks to environmental ethics as a resource for Christian ethics.

---

27 The impacts of population growth on hunger, development and the environment are much flaunted 'facts' that are nonetheless poorly understood or at times misrepresented by commentators in the transgenics debate. Cf. Sen, *Development as Freedom*. Crop scientists themselves have grown tired of the unfounded promise that transgenics appears to proffer, that it is capable of 'feeding the world'. It is an example of how a laudable aim can be hijacked for goals other than those stated. This is more than the 'empirical fallacy' at work, it is tantamount to deliberate misrepresentation. The causes of hunger are not overcome with simplistic technological promises, yet as a misconception it has shown remarkable staying power, particularly in the popular imagination. Cf. page 193.
28 Northcott, *The Environment and Christian Ethics*, 40.
29 De Schrijver, *Recent Theological Debates in Europe*, 121.

Environmental or ecological ethics?

Ecological ethics and ecotheology are terms that have been embraced by many theologians. The use of 'environmental', it is argued, resonates from the outset with too much human presumption about the non-human world, sharply segregating humanity from the rest of created reality.[30] However, following Palmer, I propose to use the term 'environmental ethics' for two reasons, one descriptive, the other normative. Firstly, 'ecology' is too restrictive in the empirical sense. It developed in the natural sciences from studies of population dynamics. As it is understood in the biological sciences it is the study of the relationships between living creatures in their biosphere. In contrast, I would suggest, the environmental sciences have a broader remit than ecology, since they are concerned with more than population dynamics and systematic interdependence. Environmental science does not insist on always taking the 'holistic' route in methodological questions; it has a wider empirical base that considers not just dynamics and interrelatedness but also issues at the level of singular entities, individual species and even individual organisms. The intention in using 'environmental' sciences and not ecology is to value the painstaking scholarship and standing of analytical scientific methods. Environmental science provides a wider knowledge-base for ethical evaluation and policy formation than ecology.

Of course ecology is used not just in the biological sense but also in the anthropological sense; as the study of the relationships between humans, as cultural and technical beings, and the world of nature.[31] As the dogmatic theologian Alexandre Ganoczy suggests, it is in this latter sense that ecology and environmental questions are the object of dogmatic and ethical reflection in philosophy and theology. Given that that is the case, it is hard to avoid the possibility that 'environmental' may be less ideologically loaded than 'ecological'. Palmer suggests that the central concern of environmental ethics is to contribute to the 'construction of

---

30 Palmer, 'A Bibliographical Essay on Environmental Ethics'.
31 Alexandre Ganoczy, 'Ecology', in *Handbook of Catholic Theology*, ed. W. Beinert and F. Schüssler Fiorenza (New York: Crossroads, 1995).

a coherent ethical system which can attribute value both to individuals and to species and ecosystems, and can also provide some way to resolve situations of ethical conflict between them'.[32] What this implies is that while the tensions between biological and normative descriptions are maintained in environmental ethics, they are more likely to be levelled in ecology. For these reasons – that ecology has a descriptive limitation and a normative limitation – the terms 'environmental' theology, philosophy and ethics will be used here rather than ecotheology, ecophilosophy and ecoethics. These preferred terms, it is hoped, will not unduly prejudice the debate on how to conceptualise the relationship between the human and non-human world.[33] It does, however, imply a clear distinction between nature and culture and a critique of attempts to subordinate the human person to a monistic concept of nature. I now turn to the three typological models that map the field of environmental theology, anthropocentrism, ecocentrism and theocentrism.

Anthropocentrism in theological perspective

A core identifier of this position is the privileged role the human person plays in creation as God's steward on earth. This is fundamental to Christian anthropology and is a category shared by all three Abrahamic faiths. Anthropocentrism, as regards non-human creation, can encompass a surprising diversity of interpretations within certain limits. At one extreme anthropocentrism can be interpreted as a kind of anthroposolipsism; where other creatures are valued solely for their benefit to humans. As a truncated view of the non-human creation this is not a position commonly defended by Christian theologians. It is however a perspective that is often implicit in economistic utilitarianism where human actions are evaluated in terms of costs and benefits, and where natural resources are presumed to be goods that are commensurate and substitutable.

---

32 Palmer, 'A Bibliographical Essay on Environmental Ethics', 90.
33 Ibid., 68.

However, anthropocentrism can also recognise a non-instrumental value for 'nature', in contrast to utilitarianism, that is also compatible with an insistence on the incomparable dignity of the human person.[34] In policy terms, as we have seen in chapter one, an appreciation for the non-instrumental value of natural resources has encouraged environmental managers to consider economic 'disutilities', in the calculation of cost and benefits; disutilities such as aesthetic and recreational value. Such considerations, although outwardly consequentialist, have an implicit 'deontological' aspect to them – both in the sense of recognising some limits to action and in the transcendental sense of seeing outer-nature as a 'condition for the possibility' of human action. Nonetheless anthropocentrism stops short of extending equal status with humans to either members of other species or to biotic communities. It specifically avoids any hypostatisation of species or ecosystems to that end.

Critiques of instrumentalism

Where it is uncritically applauded, theological anthropocentrism has several ethical shortcomings. The assumption that the human person can always improve on what has already been given appears, from an ethical perspective in theology, to be carelessly 'optimistic' and 'progressive'. It presumes that science, as a human activity, serves mankind and in doing so serves God. It keeps faith with the pioneer of the inductive scientific method, Francis Bacon, who had a clear goal in his systematising of scientific procedure, namely to give mankind mastery over the forces of nature. However, a misplaced confidence in human technology can offer few objections or little resistance to the costs of the harmonious marriage of science and Christian faith. Anthropocentrism can overestimate the human capacity for dominion, and too-readily elevates the human person to the status of 'co-creator' – a status that the biblical injunction at no point confers, and a claim that will be investigated below.

---

34   Cf. Margaret Atkins, 'Flawed Beauty and Wise Use: Conservation and the Christian Tradition', *Studies in Christian Ethics; Ethics and Ecology* 7, no. 1 (1994), 4–5.

Anthropocentric positions can also be taken to task for not registering insights from feminist theologies, which base their environmental theology, and their accounts of God and nature, on a critique of patriarchy. Feminist critiques give a corrected picture of the sort of natural essentialism that associates woman with nature and then proceeds to construe that nature positively or negatively. What is significant for this study is that feminist perspectives distinguish human knowledge and morality from physical determinism while at the same time they do not devalue the positive aspects of physicality.

The suggestion that the animal, natural, embodied world can be associated with the unrestrained, the primitive and the female is a characterisation of women and nature that has been resisted by feminism on two counts. The association of women with the natural, in some sense meaning the pre-cognitive or unreflective, attributes an essential nature to women which, feminists suggest, has more to do with patriarchy than any concept of natural kind. For Lisa Cahill the full humanity of women is a critical norm[35] and is at stake in these deceptive associations. Definitions of women that are articulated, justified and institutionalised on the basis of so-called innate characteristics, whether construed positively or not, amount to the denial of the full humanity of women.[36] There is a degree of biological determinism in the human condition, due in part to physicality. Yet, philosophically and theologically, we differentiate humanity from the rest of nature not on the basis of physical continuity but on the basis of a cognitive and moral discontinuity, on the capacity for self-reflection and transcendence. Feminism, in arguing for the full humanity of women, is at the same time reaffirming the particular dignity of the human person in creation. At the same time, feminism wants to avoid perpetuating negative evaluations of the 'natural' in that it does not want to deny the particularity of bodily or social experience.[37] In defending

---

35   Lisa Sowle Cahill, 'Grisez on Sex and Gender', in *The Revival of Natural Law*, ed. Nigel Biggar and Rufus Black (Aldershot: Ashgate, 2000), 242–61.
36   Ibid., 243.
37   Anne Clifford, 'Anthropology: Contemporary Issues', ed. W. Beinert and F. Schüssler Fiorenza (New York: Crossroads, 1995), 19.

the full humanity of women, it also defends those characteristics that patriarchy negatively associated with the female – spontaneity, embodiment and labouring in the service of human needs – as aspects of human experience that cannot be discounted.

Ecocentrism and body theology

Ecocentrism argues for a non-hierarchical system that does not differentiate the value and status of the human person, or the entire human species, from that of other living species (biocentrism) or whole ecosystems (strictly ecocentrism). It is the contrary position to anthropocentrism and a confrontational challenge to its dominance in environmental ethics.[38] It has established a new philosophical paradigm which has remained very influential both in North America and the Western world. Ecocentrism emphasises that human beings are intimately connected to all living things, that life exhibits profound interdependence, that each organism has its own good or place in the web of life, and that humans need not be considered as inherently superior to other species and their ecosystems.[39] It makes a principle of the interdependence that is found in ecological descriptions of the biosphere.

'Deep ecology' sees the fact of biological interconnectedness as a model for a deeper interconnectedness that permeates the entire physical realm.[40] This approach holds that we should sustain the widest possible identification with the universe because of this one function – its interconnectedness – and that this is a sufficient end in itself. Radical ecology constantly appeals to ecological science while at the same it could be critiqued for using an outdated concept of nature. It is positively evaluated as 'subversive' of the reductionist method of modern science yet, ironically, it claims for itself an ability to integrate scientific findings

---

38   Cf. Northcott, *The Environment and Christian Ethics*, 147–48.
39   Celia Deane-Drummond, *Theology and Biotechnology* (London: Cassell, 1997), 55.
40   Cf. Freya Matthews, *The Ecological Self* (London: Routledge, 1991).

in a more holistic manner. It is consequently, 'a battlefield of competing paradigms and research programmes'.[41] From the perspective of an autonomy approach, it also remains suspect where it pitches the human person against 'nature'. Nevertheless, as a 'consciousness raising' exercise it draws attention to the neglect of 'embodiment' in philosophical and theological anthropologies.

Ecocentric revivals of body theology

Environmental theology has prompted a renewed interest in the philosophy and theology of 'embodiment' in light of critiques of anthropocentrism. Christianity is thought of as having had a difficult and ambiguous relationship with the 'body' from its beginnings despite its anti-Gnostic orientation that rejected dualisms that disparaged the human body.[42] Clearly, the analysis of 'embodiment' is not new in either philosophy or in theology. Indeed Christian theology, as an incarnational theology, is in a general sense well placed to proclaim itself a body theology.[43] However, body theology or theologies of 'embodiment' have come to carry a more generalised meaning than specific reference to the incarnation of God in Jesus Christ. They deal, most commonly, with responses to questions related to social and moral attitudes towards the body in sexual ethics. However, they are relevant to environmental concerns because they thematise the creation in which we live and work as embodied persons; a theme that is also a task for environmental theologies.

Isherwood and Stuart suggest that three new developments in Christian theology appear to have made body theology (re)thinkable,

---

41 Josef Keulartz, *Struggle for Nature: A Critique of Radical Ecology* (London: Routledge, 1998), 2.
42 L. Isherwood and E. Stuart, *Introducing Body Theology* (Sheffield: Academic Press, 1998), 71.
43 Margaret Miles, 'Body Theology', in *A New Dictionary of Christian Theology*, ed. Alan Richardson and John Bowden (London: SCM, 1983), 77.

namely process thought, liberation theology and feminist theology.[44] These, it is suggested, worked together to open up new possibilities in theological thinking and destabilise dualist hierarchies in patriarchal thinking and other body-subordinating attitudes.[45] In relation to non-human nature for example the feminist theologian Sally McFague suggests that we should love nature because of the incarnation.[46] 'If God is always incarnate, then Christians should attend to the model of the world as God's body'.[47] She places an emphasis on the continuity (admittedly not the identity) between God and nature and between nature and the human person. What she advocates is a functional cosmology that allows us to know and love God through nature, both human and non-human.

It is clear that the tradition needs to be probed further on the theme of embodiment. However, Lisa Sowle Cahill questions what might be intended by this recovery of its significance. There is a sense after all in which ethics is always at some level about the body. Admittedly, she suggests, theologies of embodiment can serve 'to counteract a dualism about body and mind in which the body tends to come off as the inferior partner in an uneasy relationship'.[48] However, body theology is too ready to read all dualisms of mind and body in a negative light, rather than reflect on the anthropological significance of physicality, alongside the cognitive and moral aspects of the human condition.

Cahill points out that there is an implicit assumption that dualisms are automatically bad and monisms are good:

> Western authors writing today ... are almost unanimously inclined to see dualisms as bad and integration as a value, and to affirm that the body's contribution to

---

44 I do not propose to discuss the characteristics of, or tension between, these three disparate developments.
45 Isherwood and Stuart, *Introducing Body Theology*, 33.
46 Sallie McFague, *Super, Natural Christians; How We Should Love Nature* (London: SCM, 1997), 26.
47 Sallie McFague, 'The World as God's Body', *Concilium: Body and Religion* 2002, no. 2 (2002), 50.
48 Lisa Sowle Cahill, '"Embodiment" and Moral Critique', in *Theological Issues in Bioethics*, ed. N. Messer (London: Darton, Longman and Todd, 2002), 85.

selfhood is not only essential but is a component of the highest levels of human value and accomplishment, such as love, friendship, moral insight, and art.[49]

The social significance of the body however is not so easily read. The body is both a symbol of how, and the medium through which, social relationships are realised.[50] The biological givenness of the body may not change, but how bodiliness has been experienced and interpreted can change. Christian insights into the significance of the body, Cahill argues, have been misappropriated but have consistently rejected excessively dualist or negative perspectives of the body.[51] In both the life and ministry of Jesus the significance of the body is like a microcosm of the social order. The 'body' acts as a symbol of community; asceticism could be read less as a renunciation of the body than as a form of social resistance. It is this thesis on embodiment that Cahill leaves us with; that the human body has social and moral significance in the way that it is interpreted. Body theology points not so much to the inadequacies of our theology as the inadequacies of our anthropology.[52] Christologies are limited not by our lack of knowledge about God, which is accepted, but by our lack of knowledge of what it means to be human. The drive to redevelop body theologies shows the lacunae in Christian anthropology and Christology.[53]

---

49 Ibid., 91.
50 She is relying on the insights of Mary Douglas and Michel Foucault who, however, differ in a significant respect. Douglas sees the body's physicality as a universal to be socially interpreted, a process that is not always oppressive; while Foucault sees it as itself a social construction for the purpose of control and domination. Cf. Ibid., 91–93.
51 Cf. Ibid., 94–98.
52 Cristina Vanin, 'The Significance of the Incarnation for Ecological Theology: A Challenging Approach', *Ecotheology* 6.1/6.2 (2001), 113.
53 It would also be a danger for Christology, to suggest that the Christ should not be confined to the divine and human as coming together in the person of Jesus. This would amount to an attempt to divinise the human person or nature in order to express their significance as part of God's creation, a confusing move to resacralise nature using the equation of the God-man. There are dangers in suggesting that 'the immanence of God inherent in this model sees everyone and everything as a potential sacrament of God'. Cf. Ibid., 117.

If, as it is suggested, 'the human experience of embodiment is complex, ambiguous and diverse'[54] what can body theology conceptualise that is currently lacking in environmental theology? In common with ecocentrism, indeed as an aspect of it, it does highlight the continuity between the human person and the rest of 'nature'. In that regard it has a role to play in anthropology; in the construction of a new anthropology.[55] Lisa Cahill is correct to suggest that body theology contributes to an adequate anthropology by providing an expanded integrative view of the self that is both embodied and intrinsically social.[56]

Summary and critique of ecocentrism

Ecocentrism, in some of its radical forms, is not a defensible ethical position from the perspective of an autonomy approach. Ethically it erodes the commitment to human dignity, it is monolithic in its holism, and its application could have negative consequences for the already materially marginalised. It also presents specifically theological problems since it too readily collapses the distinctions between God and the world and it makes the question of theodicy even more difficult to grapple with, engendering an aesthetic attitude to pain and evil.

From an autonomy perspective ecocentrism is at best ambivalent on the dignity and irreplacability of the human person; at worst it is nothing short of misanthropy. A radical ecologist, it seems, would rather shoot a man than a snake and would consider it immoral for a rescuer to put his or her life at risk to save another. That would only make sense in a time of scarcity of people, and clearly there is none.[57] Even in its 'shallow' forms it creates problems for ethics that arise from the implications of conferring 'autonomy' on non-human species or ecosystems and thereafter ascribing

---

54   Isherwood and Stuart, *Introducing Body Theology*, 151.
55   Ibid., 149.
56   Sowle Cahill, '"Embodiment" and Moral Critique', 85.
57   Cf. Elliot Sober, 'Philosophical Problems for Environmentalism', in *Environmental Ethics*, ed. Robert Elliot (Oxford: Oxford University Press, 1995), 241.

'rights' to these as 'autonomous' entities. Autonomy, in Kant's philosophy, pertains to person, not to things.

Radical ecology uncritically evaluates holism as a positive attribute of all ecological systems.[58] Yet, holism can involve a submission of the individual to the greater whole, which itself smothers dissenting voices. It is also questionable whether holism can deliver on a greater respect for the environment than either reductionism or individualism. Deep ecology is as 'monolithic as the most single-minded individualism', the only difference being that the unit of value is assumed to be at a higher level of organisation.'[59] We cannot know if nature values itself. 'Meanwhile we are left completely in the dark as to the true essence of nature and confronted time and again with the now familiar litany of mantras about the whole that is greater than the sum of the parts, about balance and harmony, stability and diversity etc'.[60] We should be suspicious, therefore, of applying ecological insights to social policy.

Ethicists might even be suspicious when environmentalists espouse 'discipline, sacrifice, retrenchment and massive economic reform'.[61] These may well be required in some measure. However, sweeping assumptions about policy should alert us to questions about who it is that needs to be disciplined and make sacrifices and what implications for the developing world, for example, retrenchment and massive economic reform would have. As Kaul et al. have already suggested, developing countries may have to spend non-renewable resources to secure a sustainable living standard

---

58 The picture is complicated by the animal liberation movement which is so often associated with environmentalism but which leads to a different set of outcomes, from ecology, if pursued. Animal liberationists defend the lives of individual organisms – to be free from unnecessary pain and cruelty in particular in farming, wild herd management and in research. This is contrary to the relentlessly 'wasteful' population dynamics of ecology that sees the structure of nature as a series of killings. The two cannot be brought back together again without some sort of hierarchy and a return to some explicit form of anthropocentrism.
59 Sober, 'Philosophical Problems for Environmentalism', 241.
60 Keulartz, *Struggle for Nature: A Critique of Radical Ecology*, 13.
61 Baird Callicott, *In Defense of the Land Ethics* (Albany: State University of New York Press, 1989), 58.

for their populations unless developed countries – who proportionately consume and endanger more non-renewable resources – engage in economic 'transfers' to close the incentive gap.

It is, of course, possible to condemn ecocentrism by definition. Radical equality among and between species is not the only possible interpretation of ecocentric concerns. Ecocentrism can also seek to define non-instrumentalist values for nature that are not contemptuous of the human person. What remains central to it is that it is founded on a concern for the non-instrumental value of non-human reality which, it claims, is not adequately catered for in anthropocentrism. From an autonomy perspective in Christian ethics however, the danger that ecocentrism will erode the protections afforded to the human person – which are under threat from more than one source – outweighs the value of its worthy critiques of instrumental reason. It potentially undermines the respect for human dignity which is central to Christian ethics, while at the same time showing no pathway to an 'environmental dividend'. Non-dualist accounts of nature do not clearly produce a greater respect for extra-human nature.[62]

Ecocentrism, understood as a holistic, also presents problems for theodicy, as Michael Northcott rightly points out. Theological ecocentrism (pantheistic monism) allows us to affirm the goodness of creation, embodiment and pleasure, in an 'aesthetic', harmonious sense' but also expects us to embrace nothingness and pain as redemptive experiences.[63] In ecocentrism, human 'creativity is the evidence of the image of God in us, and we are charged as co-creators to evoke in artistic and creative achievement the perfections of harmony which are around us in the natural order'.[64] Ignoring the dualism between creator and creation, between nature and culture, between physicality and morality, in the name of harmonisation, leaves us bereft of any way to categorise moral agency, and human frailty and failure. Without these distinctions there

---

62 Northcott, *The Environment and Christian Ethics*, 161.
63 Ibid., 155.
64 Ibid., 156.

can be no moral obligation to 'steward' the creation and no culpability for environmental abuse.

Theocentrism and environmental ethics

Theocentrism aims to overcome the dichotomy between a self-centred anthropocentrism and a misanthropic ecocentrism by finding a path to 'relate to all things in a manner appropriate to their relationship with God'.[65] It

> ... argues that the intrinsic value of creation is established by its original and ongoing relationality to the creator God who loves all the objects of creation, from stars to starfish, who gifts the world to all living creatures and not just to humans and whose redemptive purposes include not only human life but the earth itself.[66]

It aims to find a basis for valuing nature for its own sake and not just as a resource for humanity, but without at the same time undermining humanity's place in the 'created order'. In environmental ethics it can emphasise the creation as a sacrament of the divine, as a means of grace, although this is not exclusively a theocentric theme.[67]

In one sense all theological ethics is 'theocentric'. Indeed it would seem contradictory, in Christian ethics, to argue against 'theocentrism'. However, just as in systematic theology, one can start either with God's revelation or with the anthropological presuppositions for hearing God's word, so, too, in Christian ethics. In a Christian autonomy approach, the justification for our actions and the motivation and 'the horizon of meaning' required for its realisation, are methodologically separate. Theocentrism rightly confirms the goodness of creation and the possibility of salvation for all creation. However, the work of translating that into

---

65 Porter, *The Recovery of Virtue*, 27.
66 Michael Northcott is building on the work of James Nash, Arthur Peacock, John Habgood, Stephen Clark. Northcott, *The Environment and Christian Ethics*, 143.
67 Ibid., 144.

justifiable human actions in relation to the non-human creation remains. It is not clear what 'theocentric' adds to an already theological context. Northcott claims that it will encourage more ecologically harmonious living through a renewed ecclesiology that is specifically local and environmentally sensitive. This claim will be tested further in chapter four.

A theocentric approach for environmental ethics?

Theocentrism offers a reminder of key theological themes; God's power over and continuing preservation of creation, creation as a gift, creaturely dependence, the human person as steward and the concomitant obligation that that entails. What distinguishes it from anthropocentric approaches, in a theological context, is that it puts an emphasis on 'relationality' in the 'order of creation'. As Porter points out, it suggests that if we act out of piety we will begin to discern systems of order that sustain human life, nature, history, culture, society and the self.[68] There is some truth in that. However the question that Porter rightly puts to theocentrism, is one that has already been asked of ecocentrism. How far is it willing to sacrifice the individual for the well-being of the community? Could it, for example, condone the use of coercion to limit family size as Gustafson, a theocentrist, appears to suggest?[69] As Porter concludes theocentrism leaves us still uncertain 'as to what it means, concretely, to relate all things in a manner appropriate to their relationships to God'.[70]

From an autonomy perspective, theocentrism collapses theological and non-theological themes too readily. We are instructed to find the 'order' in nature and to live by it. This leaves little room for human freedom but also little room for a religious imagination that might be more in keeping with a doctrine of God, the creator of free human beings, who continues to preserve creation and 'make everything new'. Theocentrism is in danger of bypassing or harmonising the distinctions that are traditionally kept between human and non-human nature, and between God

68 Porter, *The Recovery of Virtue*, 25.
69 Ibid., 26.
70 Ibid., 27.

and humanity. It is not thereby an 'automatic' route to environmentally sustainable living. It can become a series of conflations: the conflation of human society with ecology – at least where 'relationality' goes unexplained – and the conflation of human freedom with 'ordered' living. It also has a tendency – since it views interdependence in nature as a mirror image of the indwelling of the divine persons in the Godhead – to derive ecological principles from Trinitarian theology. In that way it is the opposite to ecotheology, which is in danger of projecting ecological concerns onto the doctrine of God.

A theocentric ethic, that honours the spontaneity and givenness by God of creation, cannot be simply tagged on to or inserted into environmental ethics *per se*. Indeed, to engage in God-talk as a prerequisite for environmental ethics would, at first glance, reduce all philosophical, scientific, social and political discourse to either ecclesiology or interreligious dialogue. Theocentrism may not in fact help to clarify the distinctions in theological ethics that make possible a free human response to the 'gift' of creation and which keep salvation and liberation in tension. It is this 'tension in history' that makes space for a theological imagination that can re-envisage our experience and legitimation of 'nature' from an autonomy perspective.

Summary and conclusions

Anthropocentrism's strength is that it conceptualises the discontinuities which set us apart as human persons (male and female) from non-human reality. Understood theologically, it is far from devaluing non-human nature as merely instrumental to human flourishing. At its best, it keeps the tension in the nature-culture relationship in play. Feminist anthropology, in arguing for the full humanity of women, implicitly holds to the central tenet of anthropocentrism, that it is the human capacity for self-reflection and transcendence that differentiates humanity from the rest of creation. Yet, feminist critique also prompts a reconsideration of our ambivalent views of nature, through a revaluation of theological anthropology. Anthropocentrism's greatest weakness is that it has failed to take full anthropological account of the significance of embodiment and our continuity with the natural world, alongside our discontinuity.

Ecocentrism challenges the prevailing anthropocentric paradigm, in environmental philosophy and theology, to consider the non-instrumental value of 'nature'. At its best, ecocentric body theology critiques understandings of the social significance of being 'embodied persons' and contributes, in that way, to a reconstruction of the ethical and social significance of the 'natural' in Christian anthropology. However, it can be critiqued both ethically and theologically. It erodes the unique status of human dignity, escalating the hostilities between human and non-human nature. Paradoxically, it dampens rather than engenders the obligation that humanity bears for the rest of the creation. This makes it even less likely to secure non-human 'nature' than might have at first been assumed. As a kind of pantheistic monism it also creates problems for theology. In collapsing the distinction traditionally maintained between God and creation it delivers a theology that is incompatible with Christian faith in a benevolent creator.

In contrast, theocentrism does not take the route of merely assigning an 'intrinsic' value to non-human reality. It argues instead that we should relate to nature in a manner appropriate to non-human creation's relationship to God. It is by this indirect route that Christian theology can justify a non-instrumentalist reading of 'nature'. Instead of beginning with the human relationship to non-human reality, theocentrism begins with the divine relationship with all creation. This has the effect of strengthening the resolve to resist instrumentalist and reductionist attitudes to 'nature'. However, rather than always clarifying it, theocentrism also tends to bypass the distinction between the human person and creation and the reality of human existence that brings these two into harmony or conflict.

In contrast to ecocentrism, Christian ethics from an autonomy perspective prioritises human dignity in cases of environmental conflict while also recognising that non-human reality is the context for finite human freedom and dignity. In that sense it is unapologetically anthropocentric. Yet, Christian anthropocentrism must take full account of the positive insights from ecocentrism and theocentrism. Ecocentrism and theocentrism attempt to establish a non-instrumental view of the creation in the interests of protecting it from degradation and abuse. Ecocentrism takes the route of establishing an 'intrinsic' value for nature for its own sake. Theocentrism argues that the human person should relate to creation in a manner appropriate to its relationship to God. Yet, both of these avoid addressing directly the ethical role of the human person as the caretaker of creation. Ecocentrism equates it with natural processes, and theocentrism equates it with faith.

In Christian theology the role of the human person in relation to non-human reality is predicated on the biblical injunction to dominion in Genesis 1. To explore the implications of this injunction we must turn next to the reconstructions of human relationships with 'nature' that have informed Christian ethics in different historical and social contexts. In the next section I examine models of 'stewardship' that might inform an environmental ethic.

Reconstructions of 'stewardship' in theology

Biblical dominion has chiefly been interpreted in its history of reception as stewardship. The human person participates in creation by 'stewarding' the creation as God's 'representative', just as a servant takes good care of their master's property. However, in environmental ethics, biblical dominion has been understood as a justification for 'domination', most notable by Lynne White.[71] Claims to dominion, he argues, are the ideological source of our instrumentalisation of nature. Ironically, dominion has also been taken to imply, on the other hand, that the human person is more than God's steward of creation but is a 'co-creator' with God.[72] Both of these claims will be investigated here. However, neither of these interpretations

---

71   Cf. White, 'The Historical Roots of Our Ecologic Crisis'.
72   Cf. Gregory Peterson, 'The Created Co-Creator: What It Is and Is Not', *Zygon* 39, no. 4 (2004).

seems justified on the basis of the biblical text of Genesis, nor seems to be in evidence in Jesus' self-understanding of his own scriptural heritage. Rather dominion, theologically, is more correctly interpreted as 'stewardship' in the service of God.

Nevertheless, that does not necessarily imply that there is only one valid ethical model for stewardship, since stewardship like discipleship, needs to be reinterpreted for every age. Atkins, for example, reinterprets the human person's role in creation as that of a pilgrim, because Christian views of nature were considered excessively anthropocentric.[73] It is only the model of the ascetic life that can be considered an appropriate counterfoil in the current age of profligate waste of natural resources. However, the model of steward as pilgrim has its limitations, not least that it may be more appropriate to the developed world than the developing world, to the materially rich rather than poor. Stewardship can also be interpreted in the context of delivering basic human needs in the service of society and of the flourishing life. Environmental theology, for example, is concerned to show that the human person, as embodied, also 'fits' into the world and that nature is the true context for human life. Here I will examine theological reconstructions of stewardship in the interests of developing a model of stewardship that is appropriate to an autonomy perspective which seeks to reintegrate the flourishing life within the framework of moral obligation. Lastly I shall examine how this reconstruction of stewardship adds to the normative aspects of sustainability, as a concept of convergence between science and theology.

Stewardship as domination or as co-creation?

For some theologians who are concerned to champion the cause of environmental protection and conservation, stewardship still bristles with the worst excesses of anthropocentrism and human hubris. Paradoxically,

---

73   Cf. Atkins, 'Flawed Beauty and Wise Use: Conservation and the Christian Tradition'.

for others, stewardship does not go far enough in championing human creativity, innovation and technical progress. As a result those theologians have restyled stewardship as 'co-creation'. Here I would like to examine the accusation, made by some environmentalists, that stewardship amounts to domination, and the preference for co-creation among some technology-friendly theologians.

As I have already mentioned, one classic criticism of the anthropocentric position, which takes the text to task, can be found in Lynn White's essay *The Historic Roots of our Ecologic Crisis*.[74] Frequently cited, and sometimes misrepresented as utterly damning of the Judeo-Christian tradition, White argues that the arrogance towards nature exemplified in Gen. 1:28 is the Christian ideological source of our contemporary environmental problems.

> God blessed them, and God said to them, 'Be fruitful and multiply, and fill the earth and subdue it; and have dominion over the fish of the sea and over the birds of the air and over every living thing that moves upon the earth.'[75]

The dominion given to humanity came to be interpreted, for White, as domination. Instead of being caretakers of creation, a position given by God to the most self-reflective and personal beings in creation, humanity has only succeeded in exploiting what was from the first a gift. The negative anthropocentric (even androcentric) image now associated with dominion continues to imply not 'responsibility' as may have been intended but hubris and even patriarchy.[76]

There may be some justification in the accusation that the text, in the history of its reception, has been used to legitimate domination. As Sean McDonagh[77] suggests the attitude of the first settlers in the Americas

---

74 White, 'The Historical Roots of Our Ecologic Crisis'.
75 New Revised Standard Version (NRSV).
76 Elizabeth Johnson, 'Powerful Icons and Missing Pieces', in *Preserving the Creation: Environmental Theology and Ethics*, ed. K. Irwin and E. Pellegrino (Washington: Georgetown Press, 1995), 61.
77 Fr. Sean McDonagh's writings are inspired by his experience of Irish rural life and by his development work overseas. This study shares many of his concerns and

and in Australia justified taking the land of indigenous people who, they argued, not being engaged in commodity agriculture, did not take care of the land as God intended.[78] For McDonagh, this view of the human person resembles too closely that of the absentee landlord, who is ignorant and neglectful of his charges. Of course the domination he refers to was of colonised peoples, not of nature, although it was in the interests of securing their natural resources. Nevertheless, he suggests that hidden in the idea of dominion is a deep-seated hubris that humanity has the ability and integrity to rearrange nature for the better, despite the fact that this is often not borne out by the evidence on the ground.

Yet, for biblical scholars and theologians alike, the interpretation of dominion as domination is a betrayal of the gift of creation and of the creator. The interpretation of dominion as domination is not confirmed by the textual evidence. In Genesis Yahweh planted a garden for humanity and 'took man and put him in the garden of Eden to till it and keep it' (2.15), as Sean McDonagh points out. The Hebrew words used, he says, *abad* and *shamar*, have non-instrumentalist implications. *Abad* means to 'work' or 'till' but it also has overtones of service, while *shamar* means 'keep' with overtones of defending from harm. It is this image of the stewardship that underlies many commendable practices in Mosaic Law: rest on the Sabbath, respect for the land and for all creation.[79] The notion of the Jubilee year, the Sabbath year, when the land was left to itself may have been a vision that was never attained but one which reminded the Jewish people that creation was not for them alone. Biblical scholars have sought, by way of reply and with some success, to rehabilitate the word 'dominion'

---

develops similar themes, although with different emphases. Cf. Cathriona Russell, 'Review of Sean Mcdonagh, Patenting Life? Stop! Is Corporate Greed Forcing Us to Eat Genetically Engineered Food. Dublin: Dominican Publications, 2003'. *Administration* 52, no. 1 (2004).

78   McDonagh, *Passion for the Earth: The Christian Vocation to Promote Justice, Peace and the Integrity of Creation*, 132.
79   Cf. Ibid., 130.

arguing that the Hebrew word implies a model not of domination but of service in 'stewardship', an accompanying biblical theme.[80]

From a historical theological perspective, Peter Harrison traces the connection between destructive attitudes towards nature and the claim that they originated in the Christian doctrine of creation. Just as the textual evidence does not bear out this conviction, for the OT and NT periods, likewise Harrison can find no such connection in patristic and medieval accounts of dominion. These, he says, are not concerned with the exploitation of nature but rather with the allegorical interpretation of natural world. Nature was to be known in order to shed light on the moral and spiritual meaning of human existence and that of the sacred page.[81] Literal readings of human dominion over nature were subordinated to spiritual readings of self-control: the will was to dominate the wayward senses.[82] That the 'conquest of nature' was well underway in the Middle Ages, for Harrison, reflects pragmatic rather than any ideological concerns. The revolution in the world of thought and culture, he suggests, from Descartes, altered the view of the human person and the place of humanity in the cosmos. Ironically it was as confidence was being lost in the role of steward and in God's bounty that nature was reconceived as hostile and had to be forced to yield its bounty.

In theology the concept of 'stewardship' has redeemed 'dominion' from accusations of domination and exploitation. It has also claimed a beneficent role for itself in environmental ethics. In its practical and

---

80 'Stewardship' is defined as the management of a property or household by a servant on behalf of the rightful owner. Its Greek root (οικονομια and its cognates) occurs frequently in the NT. The idea of holding property 'in trust' was widely taught although the word was rarely used. It underwent a revival in seventeenth-century N. America being interpreted as financial aid for pious causes. In the twentieth century the idea of stewardship has been applied to all aspects of life, most recently the stewardship of creation. Cf. Anon, 'Stewardship', in *The Oxford Dictionary of the Christian Church*, ed. E.A. Livingstone (Oxford: Oxford University Press, 1997), 1543.
81 Cf. Peter Harrison, 'Subduing the Earth: Genesis 1, Early Modern Science, and the Exploitation of Nature', *Journal of Religion* 79 (1999), 91.
82 Ibid., 96.

imaginative potential, it has migrated from its theological context – where it was put to the service of God – to 'secular' environmental ethics where it underwrites contemporary professional codes and policies. For example in the proposed Code of Ethics for the Society of American Foresters we find the following statement; 'Serving society is a profession's primary objective, and stewardship of forests is the means by which this profession serves society'.[83] Stewardship has the positive connotation of service rather than 'power over' – inherited from the original meaning in Genesis – but here implying service in the interests of society rather than God.

Yet, just as dominion has been taken to task for aiding and abetting domination, so too stewardship has been taken to task for not emphasising enough the role of human freedom and creativity in God's ongoing 'preservation' of his creation. It has been proposed that 'dominion' might be better interpreted, not as 'stewardship', but as 'co-creation'. The concept of co-creator, it is suggested, validates the role of human freedom and could serve as a point of contact in the dialogue between science and religion. Yet, although the concept of co-creation has had a career in Christian theology, it is full of ambiguity.[84] Its implication, that the human person has an equal role with God in creation, seems more than a little presumptuous. Admittedly, there has been an attempt to redeem it for Christian theology in the elaboration of the concept of the 'created co-creator'.[85] The addition of 'created' to 'co-creator is intended as a reminder that the creation is first a gift from God and that it is only in this context that the human person can be thought of as a co-creator. For Peterson, for example, this is a theologically valid interpretation of dominion because 'it implies not simply that we are creating in and of our

---

[83] S. J. Radcliffe, 'A Professional Code of Ethics for the 21st Century', *Journal of Forestry* July (2000), 17.

[84] Cf. John W. Houck and Oliver F. Williams, eds, *Co-Creation and Capitalism: John Paul II's Laborem Exercens* (London: University Press of America, 1983). Cf. Michael S. Northcott, 'Concept Art, Clones, and Co-Creators; the Theology of Making', *Modern Theology* 21, no. 2 (2005).

[85] 'Created co-creator' is a concept put forward by Philip Hefner and explored by Peterson. Cf. Peterson, 'The Created Co-Creator: What It Is and Is Not'.

own right but that our creative acts are in co-operation with God's creative acts in a way that suggests partnership rather than subordination'.[86] Nevertheless, the fact that co-creation needs additional clarification only demonstrates the difficulties it presents for Christian theology in the first instance. In stewardship 'createdness' is both a gift and a task, and that is what is clearly intended in dominion. In the concept of co-creation only the aspect of the task appears to remain. From the perspective of autonomy, it is precisely because biblical dominion implies 'service' – maintaining the distinction between divine and human creativity, between gift and task – that it is more appropriately interpreted as stewardship rather than as co-creation.

Stewardship as pilgrimage or service?

The model of the pilgrim, living lightly and journeying towards God, was traditionally designed to encourage human self-control. It has been applied as a model in contemporary environmental ethics to protect non-human nature from the excesses of human interventions. Here I will investigate its potential and suggest that the model of service, implicit in traditional views of Christian anthropocentrism and stewardship, is more apt to address environmental issues in the contemporary context.

The model of pilgrim has been recently employed in the interests of an 'environmentally friendly approach' and I present it here as a theological interpretation of the call to 'stewardship'. For Atkins, however, it exceeds the model of 'stewardship'. She implies that Christians must go even beyond stewardship to work towards the liberation of creation. She suggests that the model of pilgrim is both less interventionist and more 'egalitarian' because it implies a less active participation with creation. The 'way of the pilgrim who travels lightly sits comfortably with the egalitarian's ... concern to *allow* other creatures space to live their own

---

86   Ibid., 3.

lives, even more so perhaps than the more interventionist approach of stewardship.'[87]

Atkins proposes an Augustinian environmental approach. Augustine's 'theocentricity gives him a model of good living that sets strict limits to the individual's use of other creatures'.[88] Ironically, it is not the 'this-worldly' but the 'otherworldly' focus of the Augustinian pilgrim that limits his material wants and in that way his 'footprint' on the environment. She also concludes that in the shift from anthropocentricity to theocentricity, there is a concomitant conceptual shift from a model of steward to one of pilgrim. The image of the steward, from the point of view of the pilgrim, is too optimistic about human capabilities. Pilgrimage rather denotes an ongoing journey towards God that enables a 'healthy Christian asceticism'. This asceticism needs to be cultivated as a virtue is cultivated; by educating and disciplining the emotions and by the practice of the Christian virtues. It is to be achieved through the practice of fasting and self-denial, even if that is punctuated by regulated celebration and festivities. In such a theological reconstruction of an environmental ethic, ascetic practices are undertaken not only in the service of God but also in the service of his creation. Undoubtedly this vision carries with it a profound trust, as well as response, to the gift of createdness.

However, the 'pilgrim path' evokes a model of stewardship that draws on the ideals of monastic living; which commonly takes concrete form in a 'rule' exemplified by a life of poverty, chastity and obedience.[89] Löser suggests that this model of discipleship was influenced by several factors, to some extent by Jesus' own experience of life, by his call to perfection in the Sermon on the Mount (Mt. 5:48), as well as the appreciation for the contemplative life in Hellenistic culture. It was also influenced in later centuries by changing political circumstances. The rule of St Benedict,

---

87   Atkins, 'Flawed Beauty and Wise Use: Conservation and the Christian Tradition', 11.
88   Ibid.
89   Werner Löser, 'Monasticism: Religious Life', in *Handbook of Catholic Theology*, ed. W. Beinert and F. Schüssler Fiorenza (New York: Crossroads, 1995), 487.

Environmental Theologies and Reconstructions of Stewardship 189

for example, gave direction in the unsettled, strife-torn Italy of the sixth century.[90]

There are, however, other models for Christian living that represent the Christian response to changing social and cultural contexts, even if the monastic ideal remains a model for emulation in terms of Christian purity.[91] Lutheranism, for example, valued all forms of discipleship as theologically equal. Indeed since the Second Vatican Council the differences between 'religious' and 'secular' ways of life have been relativised, according to Langemeyer. In addition, he suggests, that although monasticism bases itself on an ascetic interpretation of Jesus' life, it is the intention of Jesus' life and not its concrete manifestation that is crucial for discipleship. He argues that 'the decisive theological criterion for the discipleship of Christ cannot lie in the greater or lesser assumption of Jesus' concrete way of life. It is, rather, to be sought in the intention of Jesus' way of life to serve in bringing God's love for humankind and God's salvific will into the world'.[92] Discipleship has, therefore, to be interpreted in each respective age.

Although the model of the pilgrim stands as an exemplar of important counter-values in human moral identity, it is not clear that this is the most appropriate model for human relationships with non-human reality. It assumes that the human person is a stranger in the world. It also runs counter to other emphases in environmental theologies which want to thematise the continuities between the human person and nature, which view nature as the context for human freedom. As we have seen in Freyne's analysis, biblical theology also gives us other images of Jesus' response to his environment than that of the ascetic, such as sharing the bountiful creation. From an autonomy perspective, the 'pilgrim path' implies too strong a suspicion of human reason and action, even if it can provide some defences from the excesses of instrumental reason. It also faces the

90 Timothy Fry, OSB, ed., *The Rule of St Benedict* (New York: Vintage Books, 1998), xxviii.
91 Löser, 'Monasticism: Religious Life', 488.
92 Georg Langemeyer, 'Discipleship: Christian', in *Handbook of Christian Theology*, ed. W. Beinert and F. Schüssler Fiorenza (New York: Crossroads, 1995), 178.

danger, as Ireaneus observed of gnosticism, of the excessive demands of asceticism being imposed by community decree. It can, in consequence, only be chosen freely in mature adulthood.

My criticism of the pilgrim path is that it is not the only possible model for Christian living and indeed it could conflict with other duties in social and family life, the perfect and imperfect duties involved in the rearing of children for example, or the demands of social justice and engagement with the world. The steward also has to rise to the task of providing for the needs of all society, including those who cannot provide for themselves. The image of the steward, therefore, is more often associated with a positive evaluation of human creativity and obligation, than pilgrim often implies. In that context it is better interpreted as human creativity in the service of human needs, society and God, as has traditionally been the case.

Stewardship from a Christian autonomy perspective

The autonomy perspective is more positive than the pilgrim path about human rational and moral capacity. As a transcendental method developed by Immanuel Kant it takes the spontaneity and creativity of human knowledge as its starting point.[93] Yet, like Kant, it also conceptualises, in its anthropology, human finitude and contingency. It is therefore compatible with Christian concepts of stewardship that neither elevate the human person to co-creator nor reduce human creativity to domination or instrumentalisation.

As we have seen in chapter two, Mieth reconstructs the categorical imperative to capture partial perspectives on the human relationship to bodiliness and continuity with non-human nature. In environmental ethics, as we have seen, the so-called emptiness of the categorical imperative, its formality, is not a drawback but its opportunity. It opens the way

---

93 As De Schrijver points out 'transcendental' refers here to the conditions of knowledge and should not to be confused with 'transcendent' which refers to God's non-creatureliness. Cf. De Schrijver, *Recent Theological Debates in Europe*, 123.

for ethical awareness and response beyond our own ethical empirical experience, and a way to continuous revision in the face of new challenges.[94] In light of the challenge from environmental ethics, an autonomy perspective can conceptualise the continuity between the human person and of 'nature' without complete identification.

From an autonomy perspective, as creatures we are also 'natural' and as natural creatures we can be also be described philosophically as dependent in that we are 'contingent'. This sense of finitude, that was evident to Kant, is likewise retained in Mieth's theological reconstruction. In classical philosophy this has the meaning of 'the dependence of being, being time dependent, finite in the sense of ending in death, capable of mistakes, the imperfection of all that is human' (Mieth, 20). Mieth highlights that contingency conveys that we can miss the point of what it means to be human; we miss the point in 'the expectation of humanity's perfectibility, of the principal feasibility of achieving anything, and of being able to still correct all problems that are created anew by problem solving' (20). We find therefore, in Kantian philosophy, equivalents to the Christian concept of *imago Dei* and sin – respectively in the positive evaluation of spontaneous creativity and in the concept of contingency. In that sense the 'stewardship' of creation has parallels with philosophical environmental ethics.

Nonetheless, from a Christian autonomy position, it is the context of faith that provides the motivation that compels us to keep these questions alive, not just because they interest us but because faith constrains us to do so. As a Christian ethic, the autonomy approach keeps to the fore the understanding that the natural world we inhabit is the gift of a benevolent and caring God. It is this that prompts us to interpret stewardship as service, rather than as domination or co-creation. The task for Christian theology is to reconstruct stewardship so that it is in keeping with other principles in Christian ethics, and registers the problems identified by environmental theologies. It is in this context that I examine how stewardship adds to the normative aspects of the concept of sustainability. To

---

94 Ibid., 132.

this end I take up the question of the relationship between sustainability and development in Christian ethics.

Stewardship and sustainability

As I have already suggested, stewardship has a history in environmental ethics outside of theological contexts. It also occupies similar ground to other concepts used in environmental management, including sustainability.[95] As Worrell and Appleby suggest, however, stewardship builds on sustainability by including a wider view of who benefits from the management of environmental resources. From an autonomy perspective – which prioritises the irreplaceable dignity of the human person – stewardship contributes to the normative aspects of sustainability by suggesting that distributive justice and development must be built into environmental management and policy, from the beginning.

There is a general assumption that combating poverty through development is an urgent humanitarian concern.[96] Yet, environmentalists have questioned whether the aims of development, with an implicit commitment to distributive justice, are simply incompatible with environmental sustainability. In this section I would like to engage in a partial investigation of that claim, first by examining the presumed conflict between sustainability, population growth and food production, and second by examining the presumed link between development and permanent economic growth. As part of this analysis I will assess the claim that food biotechnology will contribute to sustainable development in the short and long term. I will specifically examine an interim report, by the Irish Department of Agriculture and Food on the co-existence of GM and

---

95 Richard Worrell and Michael C. Appleby, 'Stewardship of Natural Resources: Definition, Ethical and Practical Aspects', *Journal of Agricultural and Environmental Ethics* 12 (2000).
96 Georges De Schrijver, 'Combating Poverty through Development: A Mapping of Strategies', paper presented to School of Religions and Theology, T.C.D. Autumn 2006.

non-GM crops in the Irish context, in order to make explicit some of the ethical assumptions made in the development of policy in this area in the EU (2005-6). The report proposes regulatory guidelines for the marketing of transgenic food under labelling restrictions and for mandatory and voluntary measures to ensure the coexistence of GM and non-GM crops in Ireland. It also addresses the question of liability in the case of market losses to non-GM farmers through the adventitious presence or introgression of transgenes into non-GM crops. From an ethical perspective the report is unjustifiably biased in favour of the GM seed industry over indigenous farming practices. The underlying assumption in the report, that GM technology contributes to a more sustainable agriculture in the Irish context, is put in question by this ethical analysis.

Sustainability, population growth and food production

I do not propose here to define distributive justice or to comment on critiques of concepts and theories of justice in contemporary scholarship. Rather I want to investigate the claim that the goal of securing the sustainability of non-renewable natural resources over time, conflicts with the goals of development that prioritise the alleviation of social injustice, inequality and poverty.

In a much quoted paper, the environmentalist Garret Hardin claimed that the human species is a tragedy for the earth.[97] Although not precisely a misanthropist, Hardin relentlessly argued that catering for the poor, in a world of finite resources, was akin to 'flooding the lifeboat'. Echoing Malthus's prophecy, Hardin predicted that the problems of population growth would require that all ethical and moral norms be subverted and superseded. The remorseless working of things (the tragedy of inevitable population growth) will result in our common natural heritage being eroded beyond repair. In response to this tragedy Hardin recommended, as regards human reproduction, 'mutual coercion mutually agreed upon'.[98]

97 Garrett Hardin, 'The Tragedy of the Commons', *Science* 162 (1968).
98 Ibid., 1247.

This exhortation may well have contributed to the complacency that followed the institutionalisation of violations of reproductive freedoms in China since 1979.

Although Hardin's perspective seems extraordinarily naïve almost 30 years later, particularly given the work of the economist Amartya Sen, it has remarkable staying power in the popular imagination. Sen has demonstrated that there is very little evidence that world food production is falling behind world population. In fact he shows that food production per head is increasing despite a concomitant fall in world food prices, something that could be expected to act as a disincentive to production.[99]

That is not to suggest that therefore there is no starvation or hunger in the world, nor that demands for food and fuel are not a factor in hunger. What he implies rather, is that the causes of hunger and starvation in the world cannot adequately be evaluated by focusing on the balance between food and population.

> Undernourishment, starvation and famine are influenced by the working of the entire economy and society – not just food production and agricultural activities. It is crucial to take adequate note of the economic and social interdependencies that govern the incidence of hunger in the contemporary world. (162)

Famine prevention is dependent not on food production but on 'arrangements for entitlement protection' (169). Sen shows that despite even a drought and a temporary dramatic decline in food production over a vast region in India in 1973, government measures to secure the incomes of those immediately threatened successfully stayed any increased mortality or undernourishment. The threat of hunger and starvation are examples of the 'vulnerability of entitlements', not lack of food. Clearly the claim that biotechnology can feed the world's growing population is unsubstantiated from an economic perspective.

That is not to suggest that environmentalists do not need to worry about the growth in the world's population. However, the alarm may well

---

99   Sen, *Development as Freedom*, 205.

be misplaced. Although it is now over six billion for the first time in the history of the earth, there are, Sen suggests, clear signs that population growth is slowing down (210). Sen concurs with Concordet rather than Malthus, a French mathematician and Malthus's predecessor and muse. Concordet anticipated voluntary reductions in fertility rates, not simply with improvements in living conditions in the economic sense, but with improvement in social conditions (218). Sen's own work indeed shows that improvements in educational and occupational opportunities, for women in particular, significantly influence fertility rates. In addition they do so over and above coercive tactics. For example, Sen compared fertility rates in China under the one-child policy – a coercive system with negative consequences for the reproductive freedoms of Chinese citizens – with those of a comparable region, Kerala in India, which did not engage in such policies. He found that 'Kerala's birthrate of 18 per thousand is actually lower than China's 19 per thousand, and this has been achieved without any compulsion by the state' (221). Both regions had a comparable birth rate in 1979 at the start of the one-child policy, and yet the rate of decline is higher for Kerala. This he suggests, argues strongly against state interference in reproductive freedoms.[100] It argues for programmes that offer educational opportunities to girls, which can delay the age at which they become mothers.

The value of Sen's work has been recognised by non-governmental agencies, such as Concern, for example, and by international financial institutions such as the World Bank and the International Monetary Fund. Yet, proponents of transgenic technology continue to make the claim that transgenic technology is vital to feed the world's burgeoning population. Nevertheless, even if hunger is only tangentially related to food productiv-

---

100 The scale of the negative impact that coercive policies have had in China and parts of India is documented by Sen. In a controversial publication in 1992, which has withstood critique, he documented that there are 100m missing women in the world due to differential access to health care and sex-selective abortion in favour of boys. Cf.Amartya Sen, 'Missing Women-Revisited', *British Medical Journal* 327 (2003).

ity we must probe further the scientific claim that biotechnology could contribute to 'feeding the world'.

From a plant breeder's perspective, in order to contribute to such a goal, plant biotechnology would need to be directed towards improving the production of those crops that underpin the world supply of food products.[101] Historically, it can be argued, substantial improvements in crop productivity did come with the genetic improvement of plant varieties.[102] As Walsh reports, conventional breeding techniques have had dramatic successes in improving the yield of crops in developed and developing countries in the mid to late twentieth century. This success, he notes, was not without its critics, who rightly recognised that hunger and starvation were the result of social inequality and not easily overcome by technical means. Nonetheless the successes inaugurated became known as the 'Green Revolution' in agriculture.

Since breeding is a task of enormous complexity, it does appear, at first glance, that 'genetic engineering' could only help to overcome some of the limitations that plant breeders experience in developing superior cropping varieties (79). However, Walsh finds, current transformations have had little impact on crop yield; although the use of transgenic herbicide tolerant crops has resulted in a modest reduction in herbicide use in intensive agricultural systems (80). They are in addition unlikely to have future impact on crop yields, he suggests. This is because 'practically all of the truly important, performance-related crop traits have a complex pattern of inheritance involving many, perhaps hundreds if not thousands of different genes acting independently and interactively, i.e. they are polygenic traits' (80). For Walsh, the blind acceptance that GM technology can provide global 'universal solutions that transcend environmental and biological barriers' may itself constitute a threat to world crop production (81). As a plant breeder, he suggests that the seduction of plant biotechnology has been such as to erode the tried and tested methods of the traditional and complex arsenal of breeders' techniques,

---

101 Walsh, 'Genes, Plants and Mankind: A Plant Breeder's Perspective on the Reality and Potential of Plant Biotechnology', 74.
102 Ibid., 77.

rather than contributing to it as it might. Clearly the claim that biotechnology can feed the world's growing population is also unsubstantiated from a plant breeder's perspective.

Development and permanent economic growth

The concept of 'development' has undergone much redefinition since its post-war, post-colonial inauguration.[103] It has its roots, De Schrijver recounts, in a cold-war ideology that sought to further technical progress in 'third world' countries to stave off the threat of communism. Development was to be effected by the global spread of the free-market economy and poverty tackled by the 'trickle-down effect' (1). Critics of this approach to development not only observed that the 'trickle-down' effect did not succeed in tackling poverty, but claimed that international institutions (the World Bank and the IMF) were undermining democracy itself, by forcing debtor countries to cut social spending on healthcare and education, and to keep wage levels low, conditional on securing loans. Since the 1970s the question of the protection of the environment has been added to the mix. Development, seen in terms of economic growth, seemed to come into conflict with the environmental imperative to sustain natural non-renewable resources (3).

De Schrijver however, also drawing on the work of Amartya Sen, suggests that development and sustainability can in fact be compatible under certain conditions. Sen departed from the classical approach to the definition of poverty and development in a series of lectures at the World Bank, later published under the title *Development as Freedom* (9). For Sen development is not measured in terms of income but in terms of freedoms or capabilities. The expansion of the real freedoms that people enjoy is both the end and means of development.[104] Increasing prosperity is not the limiting condition for improving people's quality of life.

---

103 De Schrijver, 'Combating Poverty through Development: A Mapping of Strategies'. Further page numbers are in the text.
104 Sen, *Development as Freedom*, 36.

> There are, indeed, a lot of good things in one's life that are not directly connected with income, but depend on processes of decision-making (participation in political decisions) and on the 'opportunity people have to achieve outcomes that they value and have reason to value', such as 'the freedom to live long, or the ability to escape avoidable morbidity, or the opportunity to have worthwhile employment, or to live in peaceful and crime-free communities. These non-income variables point to opportunities that a person has excellent reasons to value and that are not strictly linked to economic prosperity.[105]

Sen, as De Schrijver observes, values the economic entitlements of citizens but does not make them either the sole objective or the means of development (11).

Development is not therefore necessarily linked with permanent economic growth. Where it is not tied to growth in the consumption of non-renewable resources, it can be compatible with the goals of environmental sustainability and of Christian stewardship. Clearly for De Schrijver, as for Sen, human freedoms and capabilities can be vastly raised, even in relatively poor countries, and despite low incomes, through the development of social services rather than increased economic growth *per se*.

Of course sustainable development is not just an 'external' issue for the 'developed world'. The global market relies on public goods, many of them environmental, that it cannot itself supply and all of these goods, whether local, national or global, tend to suffer from underprovision.'[106] In environmental ethics many of the challenges are global, yet policy is still predominantly national.

The elaboration of a concept of sustainability is not simply the remit of the natural sciences, as we have seen in chapter one. The task for Christian ethics is to bring ethical insights from Christian stewardship, understood also in a 'secular' sense as service to the human person in society, to bear on the development of the concept of sustainability in the public realm. As we have seen from the analysis of George De Schrijver,

---

105 De Schrijver, 'Combating Poverty through Development: A Mapping of Strategies', 10.
106 Kaul, Grunberg, and Stern, *Global Public Goods*, xxi.

this implies that modern theology needs to appropriate those intellectual resources that are compatible with Christian ethical principles. In the next section I will examine recent proposals for the introduction of GM crops in the Irish context. The objective is to discover, in light of the preceding analysis in ethics and in environmental theology, whether GM crops are capable of contributing to the development of a sustainable agriculture, under current conditions, in Ireland.

The ethical implications of transgenic cropping in the Irish, EU context

Regulation of transgenic crops in Ireland has been the focus of a recent interim report of the Irish Department of Agriculture and Food.[107] This report was drawn up by the Minister's department after consultation with stakeholders in food production and farming and with consumer groups. Its proposals operate within the agreed limits of EU directives on the regulation of the cropping and marketing and labelling of transgenic crops. EU-approved crops (currently transgenic maize, beet, potato, cereals and oilseed rape) can be grown in Ireland subject to specific constraints in relation to labelling and marketing. The EU has recommended that each member state make regulatory provision for the 'co-existence' of GM and non-GM crops, under the principle of subsidiarity, should farmers wish to cultivate them. I will examine, from an ethical perspective, the implications of the recommendations in this report for Irish consumers and for Irish agriculture. The report over-justifies the industry's right to operate in the Irish context, and does so against consumer and farm interests.

The report acknowledges that the risk of introgression to non-GM crops, for consumption or seed (and in the case of some species to wild relatives) is considerable. However, the assumption is that currently approved crops represent no hazard to the consumer of Irish farming products or the Irish farming environment. The report therefore limits itself to addressing the claim that transgenic crops might have a negative

---

107 Department of Agriculture and Food, 'Co-Existence of GM and Non-GM Crops in Ireland'.

impact on the marketing and on-going cropping of non-GM crops in Ireland. The adventitious (accidental and occasional) presence of GM material in non-GM crops, above certain thresholds, could wipe out the premiums non-GM crops currently carry in Irish markets at home and abroad, whether conventionally-grown GM-free products or 'organic' products.

The mandate, therefore, of the Irish report is to develop guidelines, legally binding and/or voluntary, to achieve the lowest practical level of adventitious admixture of GM and non-GM crops and to comply with the legal obligations for labelling. As the report outlines, non-GM produce with the adventitious presence of GM contents above the maximum tolerance thresholds set out in legislation, must be labelled as containing GMOs. Coexistence is therefore concerned with crop management measures to minimise admixture and to minimise the economic impact associated with the admixture of GM and non-GM crops. The report also examines the liability implications where there is an economic loss or where damage occurs following admixture and suggests ways in which these losses can initially be compensated for. Such attempts to develop regulatory tools to protect Irish markets for producers and consumers do appear to be proactive and as such should be universally welcomed. However, the current report leaves several crucial ethical considerations unresolved in relation to consumer choice, mandatory and voluntary measures to control introgression, and the question of liability in the case that losses occur.

As regards consumers and labelling, the guidelines allow consumers to distinguish between GM and non-GM products, but only in some cases. Consumers will be unable to distinguish between conventionally or organically grown vegetable oils and GM oil for example. Labelling thresholds do not apply to plant extracts that contain no 'genetic' material. Yet, consumers not only recognise a distinction between GM and non-GM *products*, but they also wish to distinguish between GM and non-GM *production systems*. Under the current recommendations consumers will unwittingly be coerced into consuming GM products, since labelling recommendations will not require producers to label products as containing material derived from GM cropping (e.g. oils), or products that contain GM material, where that content is below designated thresholds. In addition farmers and growers of non-GM produce will lose the valuable designation 'GM Free'. It will be rendered untenable for all crops where any introgression occurs, regardless of threshold levels.

The report also recommends a mixture of mandatory and voluntary measures to protect non-GM crops from the adventitious presence of GM content. However, the report acknowledges that no mandatory measures are yet in place, and GM crop introductions are imminent. Likewise it fails to specify any sanctions for non-compliance with the regulations laid down. The report also takes a contradictory stance in relation to the testing of home-saved seed. It does recommend that non-GM growers should, given the danger of introgression from a neighbouring GM growers crop, voluntarily test their home-saved seed. This would ensure, for the non-GM grower, that the subsequent resultant crop would not exceed labelling thresholds, thereby incurring later market losses. Yet, in a move that contradicts its stated aims, the report suggests that the burden of costs, for this extra seed-testing, should be borne by the non-GM grower themselves and not by the GM grower or seed company. Clearly decreeing home-saved seed-testing as a 'voluntary' measure removes the onus on the GM farmer to compensate a non-GM neighbour for the extra cost of seed-testing. The testing of home-grown seed can be essential to the non-GM cropping system, yet through no fault of their own, non-GM growers must carry the cost of the continued security of non-GM seed.

Paradoxically some of the voluntary measures proposed in the report, intended to secure a workable coexistence, could not be put into practice and would only serve to undermine those that can. Securing real-time database information, in relation to rented-in land or crop schedules, in order to monitor introgression, is not a practical proposal. Indeed some of the information that would make this database useful, might in fact prove to be confidential. Disclosure might prejudice farmers' marketing strategies or undermine farm security.

Most telling is that the report suggests that the state underwrite the costs of transgenic introductions in anticipation of market losses due to

GM in non-GM crops.[108] Here the report goes further than giving Irish farmers choice; this represents a bias in favour of GM cropping. State bodies will already incur costs in the monitoring of introductions. This can be justified in that it has the dual function of protecting all partners in the coexistence process. However, state bodies should not be proposing to underwrite GM seed companies, in the absence of insurance cover, with a 'redress fund'. This has no such dual function. In addition, to suggest that the chief beneficiaries of a redress fund are non-GM growers, as the report does, is to institutionalise and systematise injury. Compensation for loss of markets should be paid by the GM development companies (principally), and GM growers. Routine compensation for 'economic losses' through introgression is not coexistence. It leaves the report open to the suspicion that the anticipated outcome of the introduction of trangenics is that measures for coexistence will simply become unworkable once introgression becomes the norm.

The report acknowledges that organic growing presents additional issues for coexistence. GM thresholds more stringent than EU regulation stipulations have been set by organic growers for organic crops. That they are more stringent than EU regulations is not a reason to ignore them, as the report does in relation to organic seed crops. Organic growers' defence of a 'zero tolerance' for GM contaminants in seed production requires that the specific thresholds for organic crops be set at the detectable level (currently 0.1 per cent). Liability for loss of seed crops applies therefore at the exceeded thresholds, whether set at 0.1 per cent (for organic growers) or 0.3–0.5 per cent (for conventional non-GM growers). If best practice in the production of certified seed would render the GM crop economically non-viable, that cannot be used as an excuse to operate less protective measures.

The report does adhere to the minimum requirements of EU regulations in terms of maximum labelling thresholds for produce and seed. However, the Department is under no obligation to limit itself to a consideration of only maximum levels. It is free, under the principle of

---

108 Although the report refers not to 'insurance' but to a 'redress fund'.

subsidiarity, to consider other thresholds, for example, those set by organic growers. Most significantly it is under no obligation to underwrite losses incurred from the introduction of GM crops, even under the auspices of a 'redress board'. That is the job of the biotechnology seed companies and GM growers. State agencies must uphold the spirit as well as the letter of the law, to protect consumers and farmers from deception and coercion 'from farm to fork'. Rather than supporting, in a one-sided fashion, GM cropping, over successful indigenous production systems, state agencies should be encouraging producer groups to argue for the clear segregation and labelling of products, to build consumer confidence and to attract product premiums. Comments from the Minister for Agriculture, Mary Coughlan (2006), which suggest that there is no enthusiasm for GM crops among Irish farmers, are not strongly reflected in the proposals in this report, presented to her office in late 2005. 'Sustainable' and 'organic' production may prove to be the more lucrative option for Irish agriculture in the long-run and not least because of strong consumer resistance to GM food in the EU. If that is so, the financial endorsement by state bodies of GM cropping, proposed in this report, is unjustified. It shows a clear bias in favour of powerful vested interests, over farming and consumer interests and a restrictive use of the evidence base to adjudicate between conflicting claims.

The Irish Council for Bioethics produced an opinion piece, also in 2005, entitled 'Genetically Modified Crops and Food: Threat or Opportunity for Ireland?'.[109] This was directed at legislators and the public alike. It concluded that genetically modified food is not morally objectionable in itself but introduces new risks for consumers, farmers and the environment. It advocated a cautious approach and a case-by-case assessment of the potential impact of transgenic food and cropping. It refers to the importance of the protection of individual autonomy and the concept of farmer and consumer choice as an instrument for its protection. At the same time is suggested that consumer and farmer choice

---

109 Irish Council for Bioethics, 'Genetically Modified Crops and Food: Threat or Opportunity for Ireland?' (Dublin: 2005).

are qualified in that they have to be 'balanced' with an 'eagerness to be in the vanguard of biotechnology'.[110] This is again an example of an attempt to limit the discussion to the individual-ethical level.[111] The 'autonomy' of the consumer is pitched against the 'autonomy' of the biotechnologist. It is significant that the EU has been accused, through the World Trade Organisation, of setting illegal barriers to trade (in the absence of scientific evidence for its *de facto* moratorium) in GM food by the US, Canada and Argentina between 1999 and 2003.[112]

It is clear from this analysis that the evidence base can lend itself to more than one interpretation. A broader ethical framework is required if state institutions are to credibly adjudicate in this area. Governments for example cannot just limit their efforts to questions of labelling, this will not do enough to prevent the erosion of consumer trust.[113] The concepts of 'sustainability' and 'sustainable development' go some way to providing such a framework, one that is inclusive of a concept of stewardship in Christian ethics. The introduction of transgenic cropping in the Irish context has little to offer Irish farmers in agronomic or indeed marketing terms and even less the Irish consumer. It would be difficult to describe such a move as an instance of 'sustainable development' under current conditions.

GM crops have not been grown commercially in Ireland to date (2008) and the official context has changed somewhat since these reports were written. With a change of government in the Irish republic and the election of Green Party politicians to the Dáil in 2007, the current Programme for Government (2007–2012) states that official policy is to

---

[110] Ibid., 2.
[111] Sigrid Graumann, 'Experts in Bioethics and Biopolitics', in *Bioethics in Cultural Context: Reflections on Methods and Finitude*, ed. Christoph Rehman-Sutter, Marcus Deuwell, and Dietmar Mieth (Dordrecht: Springer, 2006), 180.
[112] Irish Council for Bioethics, 'Genetically Modified Crops and Food', 12.
[113] Frans W. A. Brom, 'Food, Consumer Concerns and Trust: Food Ethics for a Globalizing Market', *Journal of Agricultural and Environmental Ethics* 12 (2000).

'negotiate the establishment of and All-Ireland GM-Free Zone'.[114] This does not imply a rejection of 'environmental technologies' but mirrors the actions of several regions in the EU who have moved to create GM-Free cropping zones within their borders. Increasing industry pressure is also in evidence. In both the Irish and European context the complex debate over GM food remains a debate that puts a challenging spotlight on our democratic institutions.

Summary and Conclusions

In environmental theology I have argued for retaining the anthropocentric presumption in Christian ethics. The contrasting position of ecocentrism is difficult to endorse for environmental, philosophical and theological reasons. It may be a legitimate corrective to an over-inflated anthropocentrism, and an overly reductionist methodology in science, but it shows no pathway between eroding protections for the human person and environmental sustainability. Indeed the opposite might be the case.

A 'theocentric', creation-centred approach, provides a worthy critique of anthropocentrism where it confronts the instrumentalism of economistic, utilitarian and excessively bureaucratic thinking. It places a high value on the spontaneous 'givenness' and goodness of created reality. This is not just a commonplace. A self-serving anthropocentrism that ignores human embeddedness in nature is incompatible with current Christian interpretations of creation theology. Christian anthropology is profoundly compatible with a respect for the natural world that goes far beyond a utilitarian attitude to nature as a resource for human use or abuse. Yet it also steers clear of radical ecocentrism because Christian anthropology

---

114 http://www.greenparty.ie/government/agreed_programme_for_government (18th August 2008).

recognises the central status of the human person as created in the 'image and likeness' of the God, and as the steward of God's creation.

The autonomy position unapologetically argues for the priority and the particular dignity of the human person. It affirms the positive expectation of stewardship, that the human person will cultivate what is given, rather than abuse and exploit it. It does so, however, while also defending the goodness of creation as a gift, and the possibility of liberation, redemption and salvation, for all creation, human and non-human. The promise of liberation and redemption springs from Jesus' self-understanding of his own inherited scriptures. It is through an expanded anthropology that this 'givenness' and 'goodness' needs to be spelt out in terms of environmental policy, not through ecocentrism or even theocentrism. A levelling of the traditional hierarchy – advocated in ecocentrism – only serves to escalate the hostilities and the competition and conflict between the good of the human person and that of non-human nature. A realignment towards theocentrism, in contrast, displaces the human person as steward, leaving humanity too easily off the hook in relation to the liberation of the non-human creation.

Christian models for environmental ethics have been retrieved from and reconstructed through a framework that is firmly anthropocentric.[115] What is at stake is an adequate anthropology in both our ecology and our theology. Christian 'stewardship' can be modelled on Christian asceticism; where the human person is envisaged as a 'pilgrim', oriented to a goal beyond this world, who relies constantly on the providential care of a bountiful creation, and lives 'lightly' on the planet. However, I have argued 'stewardship' may be more adequately conceived of in terms of service, where a greater confidence is placed in the human capacity for the good. In this way a Christian understanding of stewardship can contribute to the normative aspects of sustainability. As we have seen in Georges De Schrijver's analysis, one task for theology is to draw on intellectual resources compatible with Christian ethics. This is nowhere more urgent

---

[115] Benzoni, 'Thomas Aquinas and Environmental Ethics: A Reconstruction of Providence and Salvation', 446.

than in relation to global justice and global environmental issues. In the work of the economist Amartya Sen, we find a concept of freedom – as the means to and end of development – that can be reconciled with further development in the direction of a Christian understanding of human dignity and commitment to social justice. De Schrijver shows how secular intellectual resources can be positively evaluated for Christian ethics and for the development of a concept of stewardship and sustainability in environmental theology.

An autonomy approach recognises the significance of non-human nature in the development of an environmental theology and in its reconstruction of stewardship, not by personifying nature but by developing an anthropology that is inclusive of non-human reality. It supports an anthropology that appreciates human creativity while acknowledging human finitude and contingency. In environmental ethics this points towards an expanded anthropology and reads non-human nature not as external to human morality but as intrinsic to human anthropology. The concept of sustainability needs development along both descriptive and normative lines. It is the task of the natural science and economics to describe the possibilities for new technologies. In contrast it is a task for ethics, to outline the value frameworks through which our moral and ethical obligations can be both conceptualised and met. In relation to the introduction of transgenic crops to Irish agriculture, the modest agronomic advantages they promise are not sufficient to warrant their adoption, particularly not under the rubric of 'sustainable development'.

In chapter four I will explore how themes from systematic theology continue to be rich resources for a Christian ethics of the environment. In this section I have argued that 'nature' is a true context for human life, one in which we are confirmed as 'finite freedom'. It can be both an entrusted space for cultivation, and a place to refrain from action. In the next chapter I will investigate how systematic theology envisages nature as creation. The ongoing potential in Christian conceptualisations of human and non-human reality are essential to the continuous renewal of Christian morality and ethics.

CHAPTER 4

Nature in Theological Perspective

Environmental theologies have their foundations in systematic theology. In chapter three I examined the typological and heuristic models that theologians have drawn on in developing a Christian environmental ethics. I have argued for a reconstructed anthropology of stewardship that evaluates human reason and freedom more positively than some ecocentric and theocentric approaches, but is nonetheless critical of only seeing instrumental value in human and non-human nature. Ethically this reconstruction is predicated on 'nature' being the true context for finite human freedom. Theologically it is predicated on the special dignity of the human person, as a creature made in the image of God. It implies that the human person, as the one creature endowed with the freedom to respond to God's call, is at the same time at home in created reality, and is not destined for liberation and redemption from it, but with it.

Environmental theologies and reconstructions of stewardship have tended to concentrate on the explication, or critique, of two important biblical themes in their search for Christian theological models for environmental living. The first is the biblical injunction to dominion in Genesis, the second is the validation of the divine order in creation testified to in the Wisdom literature. Yet there is an inescapable tension between this two perspectives. The first demythologises the cosmos, radically distinguishing all of nature from God, while giving a specific role to human responsibility and creativity. The second emphasises the divinely created order in the universe to which humanity should conform.

Christian anthropocentrism in environmental ethics is often grounded in the first strand in creation theology, as we have already seen, interpreting 'dominion' as stewardship. In contrast, critiques of instrumentalist anthropocentrism often draw on the second strand, retrieving themes besides dominion and stewardship, such as Sabbath and Jubilee rest. The biblical injunction to dominion, it is argued – from both an ecocentric secular ethical perspective, and from a theocentric Christian perspective – results in an excessively instrumentalist and reductionist view of nature. However, both ecocentrism and theocentrism present problems from an autonomy perspective in Christian ethics. Ecocentrism tends to level the distinctions made between God and creation and between the human and non-human creation. It argues that the human person has no particular standing in creation. Ironically, this implies that the human person can also have no particular obligation for the wellbeing of 'nature'. In contrast, from a theocentric perspective, the non-human creation can only regain a non-instrumental status indirectly, through a renewal of human relationships with God. Admittedly, in theocentrism, the prior goodness of all creation is affirmed. However, the theological significance of the autonomy of created structures and created freedom, and the human person's responsibility for creation are muffled.

In recent Christian ethics, environmental theologies are most often critiqued and reconstructed from the perspective of the second strand in biblical theology – the wisdom tradition. The wisdom tradition testifies to the divinely ordered regularity already in creation. It implies that there is order in creation – alongside the directly revealed will of God – to which human knowledge and morality must 'conform'. Theologians, most notably Jürgen Moltmann, have emphasised biblical traditions other than stewardship, such as Sabbath and Jubilee rest, in order to recontextualise the call to dominion. Sabbath and Jubilee (Lev. 25. 1–55) highlight seemingly less ambivalent aspects of the Jewish tradition – a reminder that human capability and the land are the gifts of a gracious God – even if they were ideals for both the community and its land that were more aspirational than legally attainable and were the expression of a religious ideal rather than a humanitarian or ecological measure.[1] Of course Sabbath and Jubilee rest carry with them as a corollary the necessity, if not the sanctification, of human labour in meeting basic

---

[1] Freyne, *Jesus, a Jewish Galilean*, 32.

needs at other times. It seems that the human person cannot escape the responsibility of dominion.

In the Wisdom tradition of Proverbs, Sirach (Ecclesiasticus) and Job, creation is seen as a revelation of divine wisdom.[2] For Israel's sages, it seems, nature reflected a coherent order but also a coherent moral purpose, 'one in which even the ant participates in some way'. The emphasis on wisdom and prior order has the advantage at least of resisting the hubris of human claims to 'reorder nature', since it prompts us first to recognise the 'givenness' of the creation that is not of human making. It also legitimates the knowledge and wisdom of the sage who interprets the work of God in all creation. It does appear that the wisdom tradition has much to offer environmental theology: an appreciation for the order in creation that is prior to human making; a resacralisation of creation based on awe and wonder; a role for the sage (or early scientist) who contemplates and interprets that revelation; and alternative models to dominion that also sanction rest, for all creation, from human activity.

Yet the tension in the text remains, between the sacramentality of all creation and the divinely ordained role for the human person in creation; between the 'order' in creation attested to by the sages and the understanding that God 'makes everything new' in creating and redeeming. Here is not the place to examine the differences in attitude between the wisdom outlook of Hellenistic Judaism and the Genesis tradition. However, it should be clear that the concept of the God of creation and loyalty in Genesis, leads to an active responsive concept of *imago Dei*, rather than one of a passive observer of the cyclical processes of nature. Being made in the image of God has an anthropological significance that Christians would not like to downplay, even if it needs to be tempered by other perspectives. Undoubtedly it is a doctrine that is too easily co-opted for less than legitimate claims and utopian dreams. As Junker-Kenny remarks, when:

> ... the theological determination of humans as being made in the image of God is interpreted as giving free reign to us as God's 'collaborators', without any more

2   Cf. Hiebert, 'Re-Imaging Nature', 44.

definite qualifications than the need to be 'responsible', the *imago Dei* designation becomes virtually indistinguishable from a scientistic euphoric concept of human potential that is free of fallibility and finitude. If this conviction is further coupled with a 'dynamic concept of creation', the possible resistance to a market-led thrust towards a dream of perfection at the expense of the imperfect ... evaporates.[3]

Yet, if even the ant participates, how does the human person participate in God's preservation of his creation? We are left in a theological bind. If the role of the steward is to conform to a prior order, then there is little room for human freedom. If the order is such that we can never derogate from it, then there is no necessity to talk of moral freedom, culpability and responsibility. Likewise, if non-human reality has no 'function' in creation other than to serve the needs and ever-growing demands of ingenious humanity, then no resistance can be offered to instrumentalist interpretations of dominion.

In this section I will investigate some systematic foundations for an environmental theology by examining the tensions in Christian creation theology. To that end I will outline a contemporary Christian doctrine of creation that draws on the work of Wolfhart Pannenberg.[4] Pannenberg presents an in-depth analysis of historical and systematic issues in relation to the doctrine of creation. Three aspects are directly relevant here to environmental theology: the significance of the Christian insistence that creation is a free act of God; the theological interpretation of relations in the world of creatures, both human and non-human; and finally the Christian understanding of salvation for all creation.

Pannenberg's creation theology is 'mindful of human freedom'[5] and is compatible with an autonomy approach in ethics in that it develops the concepts of both order and contingency in creation, fellowship as the highest destiny (and not morality as such), and the need for redemp-

---

[3] Junker-Kenny, 'Valuing the Priceless', 52.
[4] I will also selectively draw on the work of Alexandre Ganoczy and Christiaan Mostert.
[5] Alexandre Ganoczy, 'Creation; Theology Of', in *Handbook of Catholic Theology*, ed. Wolfgang Beinert and Francis Schüssler Fiorenza (New York: Crossroads, 1995), 145.

tion. It is particularly relevant to this study that Pannenberg thematises, as he aptly calls it, the time-bound world of all creatures, not just human creatures, in their interdependence and independence. This analysis helps to make explicit the presuppositions at work in environmental theologies and it provides insights into how nature as creation is constructed from a systematic perspective. The intention is to do justice to the inner coherence of Christian doctrine first before applying it in ethics. I will then test, against this systematic backdrop, how two environmental theologians, Michael Northcott and Celia Deane-Drummond, have interpreted and appropriated some of these aspects of creation theology in developing their Christian environmental ethics. This critique, it is hoped, will retrieve aspects of creation theology that environmental theologians have had a tendency to neglect and will thereby contribute to a reconstruction of stewardship for environmental theology from an autonomy perspective in Christian ethics.

Nature as creation in the systematic theology of Wolfhart Pannenberg

Pannenberg's systematic theology makes clear the tensions in the doctrine of creation, between the call to dominion that is predicated on the human person as being made in the image of God and the givenness, goodness and prior order in creation that are not just at the disposal of humanity. Creation theology teaches that there is a given prior order to be acknowledged and protected. At the same time, God's creative activity is not limited to this original order, but God does something new in his creation. This 'novum' is revealed unambiguously for the first time, for Christians, in the incarnation and humanity has a specific role to play, in God's preservation and governance in the time-bound world of creatures. However, the human person, made in the 'image of God', also suffers from finitude, frailty and sin. This picture is further complicated by the understanding that the non-human creation itself has not escaped the effects of fallenness. Nevertheless, the human person has as an active role to play in the transformation of given structures, one that reflects creation's origin and ultimate destiny for redemption and fellowship.

Creation as the free act of the Trinitarian God

The doctrine of creation in Christian theology, for Pannenberg, begins from the faith perspective of the reality of God and traces the existence of the world to its creator, rather than the other way around. This is in keeping with his definition of God as the 'reality that determines everything'. The 'doctrine of creation traces the existence of the world to God as its origin by moving from the reality of God to the existence of the world.'[6] This is in contrast to positions which start with the reality of the world and postulate the existence of a (benevolent) creator. Although there was no early doctrine of creation, the 'articulation of the Christian belief that the God of Jesus is creator is part of a long process of the church's struggle to make its faith intelligible'.[7] Later developments in creation theology distinguished Christian views of creation from those of the ancient and Hellenistic worlds. The doctrines of creation out of nothing and the Trinity both have significant implications for creation theology and for theological reconstructions of nature as creation.

---

6   Wolfhart Pannenberg, *Systematic Theology Volume II* (Edinburgh: T. & T. Clark, 1994), 1. First published in German in 1988.
7   Anne Clifford, 'Creation', in *Systematic Theology: Roman Catholic Perspectives*, ed. Francis Schüssler Fiorenza and John P. Galvin (Dublin: Gill and Macmillan, 1992), 196.

Creatio ex nihilo

The central doctrine of *creatio ex nihilo*[8] has implications for the history of God's benevolent relationship with the creation. It crucially implies that God creates not out of neediness but from a free expression of divine love. The creation narratives offer us therefore 'an essentially demythologised interpretation of the world. The stars and other forces of nature lose their divine or demonic character: they perform serving functions within the divinely willed natural order.'[9] Nature is, by this measure, neither divine nor demonic, it is good (Gen. 1) and awaiting salvation (Rom. 8:19ff).[10] God is also the continuing creative source of present existence. 'God, as the origin, is never merely the invisible ground of present reality, but the free creative source of the ever new and unforeseen. He is the God 'who gives life to the dead and calls into existence the things that are not' (Rom. 4:17)'.[11] The promise of salvation is therefore not a promise of mere restoration – to the original Adam or the original Garden of Eden – but of transformation. For Christians this essential character of the creator, as benevolent in creating, and transformative in his historical involvement with his creation, are unambiguously revealed for the first time in the incarnation and resurrection. So it is possible to say that God does not need the world in order to be active, or to be fulfilled, but he is active in a new way in the creation of the world and in his incarnation that has implications for the present reality and ultimate destiny of all creation (5).

---

8   For Pannenberg the doctrine of 'creation out of nothing' is a later expression of the unlimited freedom of the creative action expressed in 2 Macc. 7:28; cf. Rom. 4:17; Heb. 11:3. There is no resistance to God's creative activity to be found in Genesis (13–14).
9   Alexandre Ganoczy, 'Creation, Belief in, and Natural Science', in *Handbook of Catholic Theology*, ed. Wolfgang Beinert and Francis Schüssler Fiorenza (New York: Crossroads, 1995), 136.
10 Clifford, 'Creation', 244.
11 Wolfhart Pannenberg, 'The Appropriation of the Philosophical Concept of God as a Dogmatic Problem of Early Christian Theology', in *Basic Questions in Theology* (London: SCM Press LTD, 1971), 138. Further page numbers are in the text.

God as Trinity

The idea of God as an Ultimate Unity has its roots in the Ionian philosophy of nature; in the understanding that a boundless origin can not itself have an origin (127). In its encounter with the philosophical idea of God in the Hellenistic milieu, Jews and Christians represent the universal God of Israel, the one true God, as the same one sought by philosophy (136). 'This philosophical "monotheism" was a natural ally for the Jewish as well as for the Christian mission in the struggle against polytheistic popular belief', according to Pannenberg (137).

There were also significant differences however. For Pannenberg, the Greek concept of God was shaped by its Olympian origins; God is the origin of all that exists. This concept of God is constructed by inference back from the world. The biblical God in contrast was essentially free in relation to the world. Such a free God, he says, was beyond the scope of Greek philosophy (138). The Christian creator God therefore, as the free creative source rather than the inferred God of Greek philosophical monotheism or popular pantheism, has quite a different character. He has a trinitarian character.

In Christian creation theology, creation – the existence of a reality distinct from God – is grounded in the plurality of the divine life, in the Trinity (9). The significance of trinitarian thinking therefore, in relation to the origin of creation, is that creation does not emanate as a self-centred necessity from God because God's activity is already satisfied in God-self. This confirms the Christian understanding that God does not need creation to be God and that God is radically distinct from his creation. God is 'satisfied' in the Trinitarian relations, in the 'inward' action of God.

It also allows that from a creaturely, as opposed to divine perspective, we can say that in the 'outward' action of God, something new happened with the incarnation.

> God's reconciling action, which begins with the incarnation of the Son, is really something new as compared with the establishment of creaturely existence, though it has to do with the consummation of creatures and therefore also with the work of creation. We might say very generally that what is new is that the sequence of

Nature in Theological Perspective

the divine action, and therefore of its multiplicity, is grounded in the trinitarian plurality of the divine life. (8)

It could be said therefore that creation is the ontological ground for redemption even as the incarnation and resurrection are the epistemological confirmation of the goodness of creation, from a creaturely perspective. Creation and redemption cannot be separated, although from a creaturely perspective we treat them thus.

The transformative love of God in creating, for Pannenberg, is expressed in terms of the 'outward' or economic Trinity.[12] The love of the Father is mediated through the Son. In turn, the Son is the mediator between the Father and the creature because he is the 'origin of all that differs from the Father' as Pannenberg puts it (22).[13] If the Christian doctrine affirms that God is essentially as he is shown to us in Jesus, then Jesus is 'the basis for the distinction and independent existence of all creaturely reality' (23). In turn, if the Son is thought of as the origin of difference then the Spirit can be thought of as the creative force that enables creatures to transcend their finitude and participate in the divine life.[14] 'The Spirit is the Person in the Trinity who relates what is distinct, both within the being of God and between God and created things.'[15]

---

12   Olson suggests that Pannenberg's doctrine of the Trinity is an 'exegesis of Jesus' relationship to the Father' based on the revelation of God in Jesus' life and message rather than being based on a presupposed pre-trinitarian idea of God's unity. R. Olson, 'Wolfhart Pannenberg's Doctrine of the Trinity', *Scottish Journal of Theology* 43 (1990), 185.

13   Pannenberg distinguishes the Christian trinitarian creator from the neo-platonic and Hegelian creator. In Christianity creation is the free act of a loving Trinitarian God. It is good by its original nature and destiny. In contrast Plato saw creation as originating from the fall of souls pre-existent in the logos or divine intellect. Hegel saw creation as necessary to the Trinitarian absolute Subject; as the logical element of distinction rather than the free element of distinction. Neither of these alternative conceptions captures for Pannenberg the Christian understanding of the Son as mediator.

14   Christiaan Mostert, *God and the Future; Wolfhart Pannenberg's Eschatological Doctrine of God* (Edinburgh: T. & T. Clark, 2002), 169.

15   Ibid., 171.

In summary, the insistence on the unity of the Godhead rightly mitigates against 'polytheistic' interpretations of the Godhead, including a polytheistic interpretation of trinitarian creation theology, one that simply ascribes creation to the Father, reconciliation to the Son, and eschatological consummation to the Spirit (6). Nevertheless, from a creaturely perspective, in interpreting the 'outward' activity of God, we can say that the incarnation (of the Son) revealed that the original creation (of the Father) was destined for something more than restoration but for transcendence and transformation (through the spirit) for fellowship with God. The important implication that this carries is that created existence is contingent existence, that all creation is radically dependent on God and that the incarnation confirms both the possibility of the transformation of the original creation and the possibility of divine fellowship.

God's preservation and rule of creation

In Christian creation theology there is an abiding link between the concept of creator as originator of all that is – the one who begins and preserves the cosmos in its existence – and the concept of the creator as particular, involved with all his creatures, especially with his elected people – with the history of the people of Israel. The Biblical creation terminology carries within itself a double meaning; the beginning point of creation and God's continuing relationship with his creation in salvation history. For Pannenberg the 'question thus arises for theology whether the term "creation" may be reserved for the beginning of the world or whether we must expound it as the epitome of God's creative action in world history.' (12) God's action embraces not just creation as the story of the origin of creatures but their 'reconciliation', 'redemption' and 'consummation'. 'In a broader sense the fulfilment of the creature might be included in the concept of creation' and creation 'expresses God's relation to the world in all of its aspects' (8).

For Pannenberg, Biblical writings testify to the fact that God wills to preserve the world that He has made. In the OT this is expressed in the Noachitic covenant with every living creature (Gen. 9:8–17), in Psalms (Ps. 96:10–13) where he keeps the world intact, and in caring for every creature (Deut. 11:12–15; Ps. 104:13, 27). In the NT Jesus also expresses this in what he says, for example, of the birds and the lilies (Matt. 6: 25 and Luke 12:24) (35).[16] This means crucially that God keeps the order he has made *and* cares for each contingent being in their particularity. God's creation is not limited therefore to the originating point in time.[17] For Pannenberg God's activity is better described as creation, preservation and world governance (35).

The concept of divine co-operation in the activities of creatures has the effect of keeping in tension two ideas; the idea that creatures are not left alone by God in their activities, and that God does not act to exclude the autonomy of creatures and their possible deviation from the divine purpose (48). Since the seventeenth century it could be said that the emphasis in creation theology had shifted, from the need to make room for creaturely autonomy to preventing the autonomy of the creaturely

16 I would like to follow Mueller in his use of these terms. 'The term Old Testament does not here imply a supercessionist interpretation of the convenants attested in the Hebrew Bible. Yet this study does presuppose that the Christian Bible constitutes in a significant sense a single corpus that includes the books of the Septuagint (except 1 Ezra, 3 and 4 Maccabees, the Psalms of Solomon, and the Odes as a separate book), even some parts known only through non-Hebrew manuscripts, for example, the Wisdom of Solomon and parts of Sirach. This presupposition of the unity of the Christian biblical corpus depends on the belief that one can find an authentic interpretation of its non-New Testament books in a series of events attested by the New Testament books – namely, Jesus of Nazareth's life, death and resurrection revealing him as the Messiah. I intend the denominations "Old Testament" and "New Testament" to communicate nothing but this non-supercessionist hermeneutical commitment, which is substantially the same as the one behind Paul's usage of παλαιας διαθηκης in 2 Cor.3.14 (RSV and NRSV translate "old covenant")'. Joseph Mueller, SJ., 'Old Testament and Church Ministries in Two Ecumenical Dialogues' *International Journal of Systematic Theology* 6, no. 3 (2004), 234.

17 That God continues to create, to care for existing things is often termed God's providence. However, the Christian relation of God to his creation differs from the Hellenistic concept of providence – which can be characterised as 'fatalistic determinism' or 'capricious chance or fortune'. Cf. Thomas Halton, 'Providence', in *Encyclopedia of Early Christianity*, ed. Everett Ferguson (New York: Garland, 1999).

world from becoming total. This tendency to total autonomy was related to a mechanistic world view (associated with Descartes) which developed the thesis that God, once he created the world, no longer intervenes (49). For Pannenberg, however, the preserving work of God expressly does not impede but makes possible the independence of creatures (52). Nor does it conflict with the paradisic view of the first Genesis account that suggests that creation was complete. God's faithfulness to his original creation, that includes the possibility of a free creaturely reality destined for divine fellowship, remains (40).

So for Pannenberg, from a creaturely perspective we should read creation not just as the origin of reality but in relation to preservation and divine co-operation. 'What we say about creation simply supplements the concept of preservation, expressing the dependence that unconditionally embraces all things.' (41) We also see it in temporal sequence. We experience the 'identity in change' (51). The theological concept of world governance elaborates on the concepts of creation and preservation, it is an expression of God's faithfulness in the face of changes in created reality.[18] What is particularly significant here is that this is a dynamic view of creation that implies a creative role for creatures. World governance applies to the whole of creation and it

> ... applies especially to the relations of creatures and the opposition and conflict that arises between them in their struggles to assert and expand themselves. The world as a whole is the concern of this governance. Necessarily, then, it must deal with the relations of the parts to one another ... Every creature is itself an end in God's work of creation and therefore an end in his world governance as well. (53)

Pannenberg formulates his thesis in relation to all of creation, not just the human creation, drawing on Rom. 8:19. From this he can say that from God's perspective God creates, preserves and governs all creation and that eschatologically, divine fellowship and consummation is the goal for all

---

18   This is faithfulness to, both the order first created and recounted in Genesis, and the changing circumstances in the history of the people of Israel as recounted in Exodus.

creaturely reality (Rom. 8:19ff). World governance implies something more than preservation, namely the faithfulness of God to his creation in what, and indeed even despite what, it has become.

Clearly, from the creaturely time-bound perspective creation and consummation do not coincide. There remains a tension between the present – that we might protest often 'offers little evidence that a God of love and mercy or even justice controls it' (54) – and the hoped-for future consummation. It is not clear what direction God's preservation of creation and divine world government will take us in.

> With the tension between the present concealment and the future consummation of God's royal rule over the world, the question arises whether and in what sense the divine world government has a direction, a final structure of action. (55)

It is oriented to a final consummation but it is also a present expression of God's love and participation in the life of creatures and their time-bound structures (57). For Pannenberg God did not create in order to glorify himself, although the goal of creation may indeed be to praise God, as Aristotelian Scholasticism and the older Protestant dogmatics would hold (55). The divine act of creation (preservation and governance) for Pannenberg is better thought of first as an expression of God's love, whose content and object is the consummation of creation and creatures (56).

The preservation of creation implies that God protects the general condition of creaturely independence. Divine world governance implies that God's supremacy will be over the misuse of creaturely independence. Governance implies God can bring good even out of creaturely wickedness. 'God's skill in government shows itself in His constant ability to bring good even out of evil. Its final vindication, of course, will come only with the eschatological transformation and consummation of the world as the kingdom of God.' (59)

The idea that in Christian creation theology, creation is a free act of a loving God, is now applied to preservation. Creation refers not just to an originating point in time but the continuous participation of God in his creation. God preserves the order that he has made and governs the world in what it has become. In this way divine creation is continuous. Preservation and world governance differ from creation only as an expression of the 'participation' of God in the life of creatures and their time structures (57). Preservation links the original creation to its future consummation.

This understanding of God's relation to his creation has implications for the reality of the world of creatures in the present. In turn this has implications for Christian interpretations of the natural world of creatures, for the world described by science. In advance of the eschaton and salvation for all creation, we must examine what Pannenberg's 'expanded doctrine of God' implies for an expanded anthropology that takes seriously the goodness of creation and the time-bound world of creatures.

The time-bound world from a creaturely perspective

After anchoring creation in the free act of the Triune God, Pannenberg turns to the question of creation from the perspective of the time-bound human creature. The created reality that God preserves and governs is, from the creature's viewpoint, contingent, dependent, finite and plural. In contrast to the concept of the unity of the divine reality, the 'creation of a reality that is distinct from God, but one that God also affirms and thus allows to share in fellowship with himself, is conceivable only as a bringing forth of a world of creatures'. (61) Creaturely reality is necessarily plural. From the perspective of creatures 'plurality results logically as part of the concept of the finite' (61). In consequence of this plurality a finite being is limited by other beings and a finite being exists as a distinct being only in relation to other beings. This has implications both for the role of the human person, as a finite, contingent being in creation and for the questions of the origin of evil and fallenness in creation as well as for environmental theology, as we shall see.

Order and contingency in nature in relation to the unity and plurality of creation

There is both order and unity, as well as contingency and plurality in the time-bound world of creatures. It is significant for Pannenberg that the biblical narrative does not suppress the contingency of worldly events, as well as the order in creation (66).[19] As has already been suggested, there is a tension in the Biblical account, between the action of God in the world in the regularity of natural events and his historical action in the election of Israel. This tension, for Pannenberg, found its plainest expression in the Wisdom literature (68). There were attempts, Pannenberg says, to mediate between these in Jewish thinking. Yet, for him, the full resolution of this tension, between the given order in the world and contingent historical events, comes only with the specifically Christian doctrine of the Logos.[20] In the NT, creation is related principally to the saving significance of Jesus and in the early credal statements (1 Cor. 1:24, 30; 1 Cor. 8:6) Jesus is the wisdom of God through whom all things were made.[21] In John 1:1 Jesus is likewise reinterpreted as Wisdom, as the Word of God who was with God from eternity, through whom all things were made, the creative principle of the cosmos (209).

19 There is more that takes place in nature than the uniformity of processes would allow for. The course of time, for example, is irreversible. Yet, the contingency of events is not usually a theme for science – it is, rather, presupposed. 'The concept of contingency is marginal in the logic of scientific statements. It is a correlate of regularity and not a general characterisation. It occurs only as indeterminacy' (70).
20 It will not do, Pannenberg says, for theology to set natural laws and historical experience in antithesis. It is important instead to stress 'the contingency of events as the basic character of all occurrence'. Contingency may indeed be a philosophical formulation of the Christian idea of creation and it applies to all events and therefore to the world as a whole (69). The regularity of nature is not an argument against belief in creation. A theological and philosophical understanding of the contingent nature of reality embraces the basic theses of science (71). The concept of 'contingency' is a 'concept of convergence' for theology and the sciences.
21 Clifford, 'Creation', 208.

Pannenberg, however, reminds us that we must distinguish between the unity in creation and the order in nature as understood in the natural sciences. Scientific descriptions of reality, for example, tend to focus on the order in nature in terms of natural laws.[22] However, law is an abstraction, detached from creatures in their concreteness. Natural law abstracts from the reality of the individual; in the interests of discerning the law 'behind' observed cases, it attempts to distil the 'exemplar'. In contrast in Christian creation theology the 'Logos is not the abstract order of the world but its *concrete* order'(63). The Logos is not the model or exemplar in the mind of God but is the relation between exemplar and creature; the Logos links the already with what might be, even if this can be seen only from the perspective of redemption. This implies that nature and creation are not interchangeable terms without remainder. The order in creation therefore cannot simply be equated with the natural order. From a creaturely perspective order is detected through the laws of nature, from a divine perspective it is the Logos which brings order to all creation.

For this reason we can say that the 'incarnation cannot be an external appendix to the creation nor a mere reaction of the Creator to Adam's sin. From the very first it is the crown of God's world order, the supreme concretion of the active presence of the Logos in creation' (64). So the 'Son is the creative origin of the particularity of each creature and at the same time the concrete epitome of its varied manifestation'. Creatures are more than just another particular example of an abstract principle, they are internal to the order of creation.

Likewise, the Spirit is also both the origin of contingent events and has a part to play in lasting forms. 'The power of the future that manifests itself in the dynamic of the divine Spirit is not merely understood as the origin of the contingency of individual events. It must be seen also as the origin of lasting forms and of the unvarying regularity and reliability in the process of natural occurrence, without which there can be no lasting forms' (101). The unity and plurality in created reality is linked in the

---

22 He is not referring here to Natural Law in the normative sense.

creation, incarnation and the redemption, to the order and contingency of creaturely existence.

If the natural non-human world is destined for salvation, as the concept of logos as mediator of creation implies, then it is tied up theologically with that of humanity. Through the incarnation it became clear that humans were destined for fellowship with God by virtue of being first created in the image of a God who later reveals himself as Trinitarian in the man Jesus. Christian teaching developed the concept of the central position of humanity in the universe (in spite of the existence of other creatures and despite the possibility of the existence of other superior intelligent beings in the angels). The destiny for fellowship is revealed for the first time unambiguously in the incarnation, and the incarnation, in turn, is seen as the highest instance of creation.

> The incarnation is simply the theologically highest instance of creation, the perfect realisation of the Logos in the singularity of an individual creaturely form that is not just factually different from all others but that gives validity to all others alongside itself, yet that especially validates itself before God and all creation, assuming the limits of its own finitude by other creatures. (114)

For Pannenberg the salvation of all creation is enabled by the salvation of humanity effected through Jesus Christ.

This resolution is possible because the order in the world and contingent historical action can now to be identified with both the wisdom of God and the historical person of Jesus of Nazareth (68–69). Yet, the concerns of the Wisdom literature and the particularity of the divine Logos in Jesus are still in tension and theology still needs to spell out wisdom's role in the order of creation. 'Christian theology thus has the task of finding a place for Wisdom's concern about the divinely posited natural order in the doctrine of the mediatorship of the Logos of creation' (69). That is, between the preservation of the given natural order and the 'novum' initiated by Jesus Christ.

The world of creatures and the role of humanity in creation

Given the above, creatures are not therefore simply a means to human existence. The incarnation reveals something, not about human destiny alone, but about the nature of all the divine creation. The multiplicity in creation expresses the inexhaustible wealth of God's creative power much more than anything it says about the 'entitlements' of humanity. All creation, and each creation, acclaims the Creator just by its existence. 'The multiplicity of forms expresses the inexhaustible wealth of God's creative power. As a representation of this wealth, the world of creation lauds the Creator by its mere existence' (115). God's creative activity is not restricted but extends beyond the first account of fruitfulness. For Pannenberg this 'extension does justice to the Christian interest in the uniqueness of each individual that passes on life – more so than a restriction of creaturely co-operation in God's creative activity to the mere reproduction of existing forms' (131).

However, creatureliness is not the same as fellowship; human creative 'participation in God's creative working does not of itself mean fellowship with God and His will'. This has implications for the relationship between creaturely creativity in acclaiming the creator and the meaning of the concept of the human person being created in the image of God. This creative participation in the world of creatures has to be put in the context of the destiny for fellowship with God. It is no surprise, then, that the blessing of fruitfulness that is given to humans as well, and that is even made a matter of their own responsibility (Gen 1.28), is carefully distinguished from the human destiny of divine likeness, and subordinated to it' (131).[23]

It is the concept of divine likeness that makes the human person the crown of creation and gives humans dominion over other creatures and even the earth itself. For Pannenberg, this biblical dominion expressly refers to human technology, indeed the 'primary reference in the case of

---

23 Elsewhere he suggests that the finitude of creatures, understood as 'their distinction from God and one another, will continue in the eschatological consummation'. (95)

dominion over the earth is to agriculture. It also includes mining. The charge basically embraces a technical handling of earth's treasures' (132). Dominion gives humanity responsibility for the 'preservation' of the order of creation that is oriented towards its redemption. For Pannenberg we are co-creators with God to the degree that we are oriented to the redemption of creation.[24]

> Unlike the Genesis story we may relate this responsibility less to the order established at the beginning and see it more as responsibility for the destiny of creation in its creative development. But in this form it is truly linked to the will of the Creator and its orientation to the reconciliation and redemption of creation. (132)

Humans are involved with creation not only to preserve the given orders of creation but also to participate in the creative aspects of it.

Yet, we cannot assume too much; we cannot simply equate the image and the function of human dominion (204). Theologically the basis of personal dignity is based on our creation in the image of God (176). It is an image, however, that is a gift of God: 'that none of us has by merit, that none of us can receive from others, and that no one can take away from us' (177). It is not reason as such, nor intellectual ability but the destiny for fellowship with God (202) that theologically gives humanity dominion in the time-bound reality of creatures.

Nothwithstanding that, if humanity has a role in the redemption of creation in advance of the eschaton then we must contend with the question of fallenness and wickedness in the world of creatures and the question of the goodness of creation itself.

---

24 Gen. 1: 26–28: Then God said. 'Let us make humankind in our own image, according to our likeness; and let them have dominion over the fish of the sea, and over the birds of the air, and over the cattle, and over all wild animals of the earth, and over every creeping thing that creeps upon the earth'. So God created humankind in his image, in the image of God he created them; male and female he created them. God blessed them, and God said to them, 'Be fruitful and multiply, and fill the earth and subdue it; and have dominion over the fish of the sea and over the birds of the air and over every living thing that moves upon the earth'.

The problem of evil in God's creation

The task of the doctrine of creation is to show that the world as it offers itself to human experience is the creation of the God of the Bible, that it is the creation of a benevolent and loving God (161). The Genesis story tells us that after each act of creation the work is said to be 'good'. With the creation of the human person we hear that all creation is 'very good' and humans are viewed as the crown of creation. Humans have a special place of importance in creation and the 'goodness of all creation obviously depends on that of humans and their being in accord with the divine purpose in creation' (162). Yet, we have to examine the relationship between finitude and contingency which are aspects of creatureliness and the nature of human sin and the fallenness of all creation.

Part of Christian teaching has been that all creation is entangled in the disorder that comes from sin. However, for Pannenberg, the fact that pain and suffering are widespread among living creatures does not necessarily imply that it is all the result of sin. Finitude, he agrees, is not the same as sin, even if finitude does carry with it the possibility of sin. It is, he says, surely not enough to advance the limitations of creatures as a reason for the possibility of sin – finitude is not yet evil, it may lead to error but it is not a fall from God. 'There is some truth in the tracing back of evil, including the moral evil of sin, to the conditions of existence bound up with creatureliness. Nevertheless, it is not enough simply to advance the limitation of creatures as a reason for the possibility of evil ... the limit of finitude is not yet itself evil, and if evil came from it, then we must define evil as error and not as a fall from God' (171). From that perspective, it can be said, that not all creaturely or human suffering is a result of sin.

There is natural evil, Pannenberg suggests, that is not as a result of human sin, but is part of this cost of the development of all creaturely forms (97). Creaturely independence and interdependence is part of our finitude, creatures live off and for one another and assert themselves against each other. This brings strife: for Pannenberg the 'suffering of all corruptibility that affects all living things, and that forms the background even to joy in life, culminates in the suffering that others add to it' (172). We see this assertion in the 'ascending line of forms of life' which comes to a climax in humans: and in humans it is theologically referred to as sin (173). As part of creaturely finitude, creaturely independence and interdependence is not itself evil but it brings conflicts with it that could be a source of evil. This has implications for an environmental Christian ethic. If interdependence is constitutive of creaturely existence, as environmentalists continually emphasise, theologically this carries with it ambiguities in relation to fallenness that apply to all of creation. Limitation and finitude are aspects of creatureliness both in relation to its independent particularity and to its intersubjectivity and interdependent relationality.

In terms of the human origin of sin, Pannenberg also suggests that ultimately, even if evil has its origins in the freedom of creatures, this does not absolve the Creator of all responsibility for creation.

> To point out that evil has its origin in the freedom of creatures to decide and act is not to absolve the Creator of all responsibility for this creation of his. Although creatures may be free, even in their freedom they are still God's creatures. Concern to absolve the Creator has been a mistake in Christian theodicy. (166)

Nonetheless, there is a relationship between the misuse of human freedom, and sin and evil in creation. For Pannenberg, there is 'an important element of truth in the reference to the freedom of creatures' as the origin of sin since the 'decision to create carried with it the risk of a misuse of this creaturely freedom' (166). The 'Creator accepts the risk of sin and evil as a condition of realizing the goal of the free fellowship of the creature with himself' (167). Although it is not the objective of this study to investigate the systematic foundations for moral autonomy, the human person's destiny for free fellowship points to an important corollary for the autonomy position in ethics. Moral freedom before God is a central aspect of Christian anthropology.

Wickedness and evil are not an object of God's will, they are not necessary to the perfection of the universe, but could be considered the risk, the condition of the realising of God's purpose for creatures; the creation of free respondents. In that sense evil and wickedness – although they are not creatures – come under the divine world government that brings good out of evil (167). It is possible then to say that evil is 'costly' for God as well as for creatures.

For Pannenberg, wickedness and evil are a dangerous reality for creatures but are risked by the Creator as condition of realising the goal of the free fellowship of the creature with God (169). They may have entered creation but not as objects of God's will or creative pleasure (169). Yet God takes responsibility for them in his world governance. For Christian theology it is in the union of creation and redemption, and not its disjuncture, that a tenable answer to theodicy is possible, according to Pannenberg (164).

Summary and critique

The distinctly Christian doctrine of creation out of nothing and of a Trinitarian conceptualisation of the being of God have significant implications for a Christian view of nature as creation. Both of these doctrines contribute to the Christian understanding of creation as originating in the free will of a benevolent creator. The doctrine of 'creation out of nothing' implies that God's creative act is free and gracious, he neither needed to create, nor is his activity restricted to his original creation. It is because he created freely that we can say that God created out of love and not need. This confirms the original goodness of all creation. However, it does not limit salvation to a return to an antecedent original goodness but points to the transformation of all creation.

Creation, as an outward action of the Trinity, also reveals something of the nature of the Godhead, and the destiny of the creation for fellowship with God. From the divine perspective, the incarnation is not an afterthought but inextricably links the original creation to the possibility of its transformation in its redemption and fulfilment. It is the confirmation of the goodness of the original creation, not the ground for its goodness. This confirms that the human person, as a creature, 'fits' in creation and is destined for salvation with it.

For Pannenberg, God allows evil into creation as the price to be paid for the possibility of human freedom and the ultimate hope of salvation in the fellowship of God. Yet, God continues, in his preservation of creation and in his divine government, to subvert the effects of evil and bring good. God's preservation and governance of creation, however, are not in conflict with the possibility of the independence of creatures. On the contrary for Pannenberg, these are what make creaturely independence possible. It is legitimate for environmental theologians to champion the 'relationality' in creaturely existence but not to the exclusion of the 'independence' of creatures.

Systematic perspectives bring many enriching insights for environmental ethics: the importance of individual creatures: diversity as part of God's plan for creation: the distinction between finitude and sin; and the idea that all were created by God's will, that there can be no element of competition with pre-existent matter. Pannenberg's creation theology, with its critical insights into the history of doctrine, is taken as a presentation of these points in contemporary theology. It sheds light on those systematic insights that are commonly taken up in environmental theology.

Admittedly, from the perspective of a Christian autonomy position, further questions could be put to his anthropology. It is not always clear how seriously he takes the role of autonomy as the basis of our obligation towards the other: for example when he suggests that the state be based on acknowledging a religious foundation.[25] In the context of my study, however, the parallels with Mieth's development of autonomy are significant; order and contingency in creation, fellowship as the highest destiny (and not morality as such), and the human need for redemption. His incorporation of insights from the natural and social sciences in his creation theology, in the context of the modern consciousness of freedom, and his ecumenical standing make him a worthy conversation partner.

These systematic considerations are significant, in that the tensions in creation cannot be levelled in any environmental theology. In the next section I would like to examine how two Christian theologians have

---

25 Maureen Junker-Kenny, 'Christian Ethics and Culture', in *The Critical Spirit; Theology at the Crossroads of Faith and Culture*, ed. Andrew Pierce and Geraldine Smyth, OP (Dublin: Columba Press, 2003).

appropriated these systematic insights and tensions in the doctrine of creation – the goodness and fallenness of creation, the role of the human person, order and unity in the world of creatures, the 'newness' of God's creative activity – and how that has influenced their development of a Christian environmental ethic. Michael Northcott highlights the history of Christianity's appropriation of philosophical categories as degeneration. He prioritises the importance of the goodness of the primal natural order, and builds a natural law ethic drawing on Aquinas. However, he undervalues the role of human reason and freedom. Celia Deane-Drummond, in contrast, evaluates human creativity more positively – also drawing on Aquinas. She re-appropriates aspects of the wisdom tradition, and she develops a virtue ethic in her environmental theology. In ethical terms she implies that the biblical injunction to dominion needs to be contextualised and renewed at the level of values. From the perspective of an autonomy position, her approach can contribute to an environmental theology that considers nature as creation from both a deontological and a teleological perspective.

Nature as creation in the 'natural law' ethic of Michael Northcott

Michael Northcott seeks to restore the moral significance of the natural order that he finds at the heart of the Christian tradition. To that end he develops an evangelical reconstruction of a 'natural law' ethic. For Northcott, the original moral significance of the natural that can be found in the Hebrew scriptures and in early Christianity was lost chiefly through Christianity's compromising engagement with secular culture. He details the historical roots of the current environmental crisis and develops a 'repristinated'[26] natural law ethic to defend a return to an

---

26 Northcott, *The Environment and Christian Ethics*, 255.

environmentally harmonious agriculture. He seeks and finds evidence to suggest that human communities and indeed other species can live in their natural environment without exploitation, once conscious of their fundamental relationality. Theologically he emphasises the goodness of created primal order. As a response to the disruption of nature by human activity, he reconstructs stewardship through the biblical concept of its corollary, Sabbath rest. I will outline and critique Northcott's approach in relation to the systematic presuppositions implicit in his environmental ethics. In particular, I will examine his theological interpretation of the ecological crisis as a result of Christianity's compromise with ancient Greek philosophy and with modern culture, his use of biblical models and his reconstruction of 'stewardship' as predominately a return to the primal order in creation, and his 'repristination' of a natural law ethics.

The 'ecological crisis' as a problem of modernity

It is telling where Christian ethics locates the turning point in the human misuse of nature that is responsible for our current environmental problems. For Northcott, the crucial turning point occurred during the time of the Enclosures (of common land) in Britain and the Rhineland.[27] This development initiated fundamental changes for agriculture, and attitudes to nature underwent a transformation that Northcott traces to a secularisation process. In the European context, and later in the context of her colonies, changes in land ownership, alongside influences that can be traced to the Reformation played a major role in human exploitation of nature.

> Protestant theologians emphasised more strongly than their medieval forebears both the fallenness of nature, and its consequent fearfulness, and they treated nature as a resource created entirely for human purposes. Through its human use

27 A classic presentation of this development can be found in W. G. Hoskins, *The Making of the English Landscape* (London: Hodder and Stroughton, 1955).

and transformation by Christian peoples nature might also be gradually redeemed from the effects of the fall.²⁸

There was, Northcott suggests, an inward shift towards thinking of God's redeeming activity as happening entirely in the individual self, that produced a doctrine of creation that was more instrumentalist (and secular) than what had gone before. He links the Reformation turn – to the salvation of the individual self in its direct relationship to God – to the development of the money economy and reiterates with Max Weber that

> ... it was Protestant theological teachings which legitimated and in part inspired the new culture of mercantilism and capital, the new practices of usury, private property, the enclosure of commons and the disenfranchisement of the peasantry in the Rhinelands and in England.²⁹

The changes to agriculture were profound; this secularisation process did not leave 'natural capital' untouched. In addition the negative impact was further entrenched by developments in science. Northcott regrets the demise of premodern agriculture which he equates with the harmonious cultivation and nurture of life in all its diversity.

> It represents a set of practices and of inherited wisdom by which human life and non-human life are brought into a community of interest for a range of goals including the nurture of the landscape, the preservation of natural resources for future generations, the celebration of the cyclical nature of life, the quest for harmony between human and divine purposes, and the nurture of the soul.³⁰

---

28   Northcott, *The Environment and Christian Ethics*, 53. This echoes the by now classical, provocative and, as I have shown, for the most part exaggerated, thesis that Christianity bears a burden of guilt because of the Genesis narrative's exhortation to dominion.
29   Ibid., 54.
30   Michael S. Northcott, '"Behold I Have Set the Land before You" (Deut. 1:8) Christian Ethics, GM Foods and the Culture of Modern Farming', in *Re-Ordering Nature; Theology, Society and the New Genetics*, ed. Celia Deane-Drummond, Bronislaw Szerszynski and Robin Grove-White (Edinburgh: T. & T. Clark, 2003), 87.

This previous premodern sympathetic attitude to nature, he argues, directly contrasts with modern technical agriculture with its corrosive effect on nature and society.

> Wherever secular modernity encounters traditional culture and religion, it tends to disrupt and corrode harmonious human relations with nature, and the religious sensibilities and rituals by which these relations were traditionally sustained. This happened after the Reformation in Europe when attitudes to money, trade, land and work were secularised and it continues today in the assault of modernity on the local religions and cultures of primal peoples. In contrast the traditional religious world-view of the Middle Ages tended to preserve the natural world from excessive human interference. Religious rituals were closely tied to natural cycles and seasons, and were often perceived as contributing to the fertility and beneficence of both divine–human relations and relations between humans and their animal and plant neighbours. (84)

Although it is not at all clear that such moral harmony ever existed, nevertheless, for Northcott, it is the secularisation of Christianity in modernity, alongside mercantilism, legitimated in part by aspects of the Protestant Reformation, that are currently to blame for environmental degradation.

Cultural encounter as a history of compromise

Northcott's critique of modern science, technology and capitalism has a historical theological counterpart in his critique of Christianity's encounter with philosophical concepts of God. For him, theological appropriations of philosophical categories must be read as a history of compromise, from the early encounter with neo-Platonism in the ancient world to the later conclusions of the medieval nominalists. The original creation–redemption synthesis evident in the writings of Ireaneus, alongside the early-Christian resurrection faith became distorted, for Northcott, in the subsequent 'Romanisation and Hellenisation' of Christianity (210). An exclusive focus on redemption and a growing ambiguity about the body, under pressure from Gnosticism, almost eclipsed this synthesis, he suggests. Augustine's separation of the creation of the world from its redemption under the influence of Neoplatonism is ultimately not, for Northcott, Christocentric or pneumatological. All of these influences had the effect, he claims, of gracing the human will, while the rest of creation is left outside the divine economy of salvation (216). In Augustine he sees the seeds of later nominalism which pictures creation 'as a product not of Christological care and pneumatological providence as much as of the arbitrariness of divine will' (216). This historical decline continues until the

> ... early Christian schema of creation/redemption is gradually eclipsed under the influence of Hellenistic philosophy, and in particular of Platonism, and the good and salvific purposes of God for any part of the created order other than human souls are further eclipsed in Protestant theologies of election. (221)

This analysis of Christianity's encounter with philosophical concepts of God is problematic from the autonomy perspective in Christian ethics. As Pannenberg has already suggested, the influence of Greek thought on the Christian Apologists cannot simply be read as an adulteration of the essence of Christianity.[31] For Pannenberg, Judaism and Christianity appropriated and distinguished the biblical idea of God from the Hellenistic philosophical idea of God not just because they encountered them but for sound theological reasons.

> The debate with the philosophical question about the true form of the divine was indeed occasioned by encounter with the Hellenistic intellectual world, but is also grounded in the biblical witness to God as the universal God, pertinent not only to Israel but to all peoples.[32]

Philosophical ideas of God were not simply appropriated uncritically by the Christian Apologists. Rather, they were transformed in Christian theology in the 'critical light of the biblical idea of God': the biblical God is not exhausted by his creative function or confined to it, he is the 'free, creative source of the ever new and unforeseen' (138). This systematic

---

31   Pannenberg, 'The Appropriation of the Philosophical Concept of God', 119.
32   Ibid., 134.

consideration has implications for how we found a Christian view of nature as creation. There is a theological difference between a concept of the goodness of the creation as 'the free, living gift of the God who acts in history' and 'an essential necessity of the Platonic idea of the Good, which gives itself like the sun'.[33] It is the loving freedom of the biblical God which defends belief in the resurrection and in creation *ex nihilo* and lies behind Irenaeus's theological–philosophical synthesis (178). The insistence on the order in creation has to be tempered with the concept of God's free creativity.[34] For Pannenberg, the biblical record affirmed 'the conception that the order of creation was intended to remain as it was founded in the beginning' (109). However, he says this has to be 'connected with the idea of a continuing divine activity of creation, as suggested by other biblical passages' (110). So, far from selling out to Greek thinking Pannenberg shows how the apologists inaugurated new thought forms that changed the course of Western thinking.

Northcott's theological interpretation of the ecological crisis, and the presumed degeneration of Christianity in its encounter with ancient philosophy and in modernity, influences how he interprets theological models for creaturely relations in his environmental theology and his use of natural law in environmental ethics. It is to the ethical implications of

---

33 Ibid., 175.
34 Pannenberg also points out that the mechanistic interpretation of nature as regular and ordered has been tempered in the contemporary natural sciences with the concepts of contingency and emergence. He relates this conceptually to the continuing activity of God in creation. 'The irreversibility of time with its consequence that every new event is uniquely new and the indeterminacy of individual events according to quantum theory seemed to support the view that natural processes are basically sequences of contingent events – an assumption that is not opposed to the operation of laws in their course, because precisely in the unique sequence of contingent events regularities may occur that can be expressed in formulas of natural law. The element of necessity in the concept of natural law [Naturgesetz not Naturerecht] is not in opposition, then, to the basic contingency in all natural events, though there is also unpredictable contingency and novelty.' Contingency may be the basic character of all events and not only the exception. This gives us, he says, a different view of nature at large. Cf. Pannenberg, 'Problems between Science and Theology in the Course of Their Modern History', 110.

his application of systematic foundations. in relation to the non-human creation, that I will first turn.

Stewardship as the restoration of the order in creation

Northcott takes up and critiques appropriations of the OT biblical models of dominion, stewardship and Sabbath and interprets them through the NT narratives of incarnation and resurrection. He recovers Hebrew and Christian moral concerns for the land and for creation. These are illuminated by his ecological concerns. However, he reconstructs a one-sided model of stewardship that understands stewardship as the call to conform to the prior order in creation and nature. This given order in nature is revealed to us both theologically and ecologically and can be characterised as interdependent and relational.

Dominion, stewardship and Sabbath

The Hebrew Scriptures capture for Northcott a 'primal' respect for the order of nature that treats creation as a divine gift, although not as a subject of worship in itself.[35] He upholds the distinction made in systematic theology, between nature religion and Christian theology's demythologised view of natural reality, and is correct to suggest that 'environmentalism' does not offer a solution to the instrumentalisation of nature, at least from a theological perspective. This is because environmentalism stems from an inappropriate remythologising of the natural world.

The problem with environmentalism, for Northcott, is that it 'reflects the Romantic turn of Enlightenment philosophy and culture towards nature as the source of human purpose and meaning, in a universe devoid of divine presence or moral purposiveness. It is from the Romantic quest for aesthetic value and moral purposiveness in nature, without divine warrant, that ecological thought emerges' (85). Although he misses the

---

35 Northcott, *The Environment and Christian Ethics*, 170.

theological interests of the authors of early Romanticism, his theological rejection of ecological pantheism is a clear advance on some uncritical body theologies. He is correct to suggest that the 'new ecotheological pantheistic orthodoxy' leads either to the optimistic adulation of human progress or to deep ecological sentiment, neither of which are in keeping with Christian ethical insights (162). Pantheism also presents fundamental systematic problems for theology since it offers no ground 'for distinguishing the being of God from the life of creation, and therefore seems to offer no philosophically defensible account of moral evil' (162).

By stressing that true Christian creation theology developed in contradistinction to the animism, pantheism and later Gnosticism of the ancient world, Northcott also rightly points out that even if nature is neither divine nor salvific, this does not divest it of all moral or ethical significance. Alongside this demythologising, the concepts of the covenant relationship between God and his creation – in dominion, stewardship and Sabbath – all served to retain for the non-human creation a non-instrumental value apart from human use. 'The concept of the cosmic covenant indicates that the ancient Hebrews, like the ancient Greek philosophers, found in the natural order both ethical and ecological significance' (173). In this case, he evaluates the parallel between Greek and Christian thought more positively than before, not distinguishing the Greek concept of the cosmos from the Christian concept of creation.

Dominion, as we have seen, is predicated on the understanding that the human person, in God's stead, will steward the land as a gift from a benevolent creator in whose image the steward is made. For Northcott, stewardship is too problematic from an ecological perspective.[36] He highlights instead the concept of Sabbath rest as a sign of the goodness of creation, which is valid but must be put in relation to the complementary aspect of human labour. Sabbath rest has ecological value, Northcott argues because

> ... the most significant indication of the moral status of the land is the requirement of the Sabbath of the land. The creator rested on the Sabbath from the work of

36   Ibid., 129.

creation and in this rest the goodness and order of the cosmos are affirmed, blessed and enjoyed by the creator. (188)

Sabbath, as a sign of the covenant with God, ties in with an ecologically oriented 'land ethic' that bears responsibility for the land for both present and future generations. Northcott compares this biblical land ethic with aspects of the land ethic of contemporary environmentalists.

> ... the primal reverence of the Hebrew traditions for the goodness and wisdom of created order is cognate with the beliefs of contemporary environmental philosophers that natural order and ecological community are essential features of the identity of the human self-in-relation. (165)

In summary, Northcott suggests that Christian interpretations of dominion as stewardship are too open to compromise with the prevailing culture of the day. In contrast he highlights the Sabbath motif in order to capture the insight from systematic theology that he finds most appropriate to the current crisis: the insight that there is a fundamental prior order in creation to which the human person as steward must conform.

Resurrection as the vindication of creation

For Northcott the primal order, balance and respect to be found in the Hebrew tradition is related to the redemptive activity of God in Christ, in particular to the doctrine of the resurrection (198). Drawing on Oliver O'Donovan's *Resurrection and Moral Order,* Northcott emphasises the resurrection as the starting point for both an evangelical and a natural environmental ethic. He links the Hebrew emphasis on Sabbath rest with faith in the resurrection to develop his concept of 'nature redeemed'. Ironically, the power of the Christian God 'who brings life to the dead and calls into existence the things that are not' (Rom. 4:17), is dissipated somewhat in Northcott's one-sided environmental theology, because he privileges restoration over the transformation of nature as creation.

For Northcott 'Christianity emerged from the religion of the Hebrew Bible as a consequence of one single event – the resurrection of Jesus Christ from the dead'(199). So, although it is possible to say that the creation is fundamentally good, this is 'reaffirmed by the being of God uniquely embodied in the material creation in Christ's life, death and resurrection' (199). Without 'the resurrection, the child Jesus would not have come to be known as the saviour of the world, and without the resurrection the death of Jesus would not have been seen as the sacrifice by which the sins of the world were atoned' (200). Likewise the resurrection is significant not just for humanity but for the whole of creation, since without it all creation would remain without hope of restoration.

> Christians traditionally valued the existing and material ordering of creation because they perceived that in Jesus Christ the original creative principle through which God had caused the world to come into being – the *logos* or divine Word – had been made known as a human fleshy body, and in space and time. And further the Resurrection revealed to Christians that it was God's intention to redeem life in the body of the material creation, and not just to rescue the souls or spirits of Christians from out of the organic and substantial cosmos.[37]

The resurrection points to the eschatological restoration of creation, its redemption from the frustration of evil, and the possibility of transformation. However, it does seem that for Northcott, the authenticity of a properly theological account of nature as creation can be tested on the degree to which the resurrection is emphasised in creation theology. Borrowing from O'Donovan he says that 'the creational significance of the resurrection is central to a fully Christian and theological account, as opposed to a secular or liberal theological account, of the divinely given ethical project of human life'.[38] This could close off the link between the human hope for meaning and the transformative power of God's love, in a theological anthropology.

Significantly, what the resurrection reveals, for Northcott, is not so much the transformative aspect of divine creativity but the prior order in creation and the relational, rather than self-reflexive character of human life. For Pannenberg the resurrection also reveals the essential characteristic of the creator and his historical acts. However, Pannenberg interprets

37 Northcott, 'Ecology and Christian Ethics', 212.
38 Ibid., 215.

this essential characteristic as not only confirming the prior order in creation but inaugurating the transformative newness of God's participation in continuing creation. 'The essential characteristic of the creator and of his historical acts is unambiguously revealed for the first time by the resurrection. He is the God 'who gives life to the dead and calls into existence the things that are not' (Rom. 4:17).'[39] For Pannenberg it is this insight in particular – the freedom of God in relation to the world – that is the specifically Christian revelation in the resurrection, which is inaccessible to 'natural' accounts of creation. The creative freedom of God in Christian revelation is specifically 'inaccessible to the inferential procedure that is fundamental to philosophical theology' (138).

Of course exclusively resurrectional Christologies can be suspect, as Northcott is aware. In Christian theology the resurrection may be the epistemological point of departure but to make it the basis for Christ's divinity could risk adoptionism, as Ullrich implies. 'Christ has not become divine with the resurrection, although the resurrection proves his divinity and the truth of his teaching.'[40] Having once affirmed the resurrection, we 'fall back' on the incarnation (as Northcott must) to discover how, in the time-bound world of creatures, God's kingdom has been inaugurated. Pannenberg captures this in his 'Christology of the resurrection of Jesus,'[41] where the claim of the earthly Jesus is confirmed by the resurrection as an anticipation of the eschatological self-revelation of God.

Admittedly, Northcott recounts the order and relationality in the ministry of Jesus in the synoptic gospels. 'Jesus is portrayed as one who lives in supreme harmony with the natural order' (224). Paul, he suggests, hints at the light of nature as a moral source, even for those who lack the guidance of the Jewish legal tradition (Rom. 1: 19–20). So many passages:

---

39   Pannenberg, 'Problems between Science and Theology in the Course of Their Modern History', 138.
40   Lothar Ullrich, 'Resurrection of Jesus', in *Handbook of Catholic Theology*, ed. W. Beinert and F. Schüssler Fiorenza (New York: Crossroads, 1995), 593.
41   Cf. Ibid.

> ... affirm that the Christian ethic is not simply an ethic *for* Christians, but is rather what Oliver O'Donovan calls a 'natural ethic' because it involves an understanding of the moral order of the world, and the restoration of that order in Christ, which is the order which address all people and not just Christians. (226)

Undoubtedly this also has a this-worldly reference, in the example of Jesus' life and ministry and the continuing work of the Spirit in creation.

> Through the resurrection Christians discover that the law of sin and death has been overturned and that the possibility of moral restoration in *this* life, which Jesus announces in his preaching and demonstrates in his miracles, is brought near after his death through participation in the Spirit of the resurrected Christ.[42]

Northcott, however, fails to adequately make the link between the resurrection and his primary value of relationality in nature and creation.

Pannenberg, and indeed O'Donovan, would suggest that creation's original goodness in its independence and interdependence is already established. The resurrection is the epistemological ground and not the ontological ground for the goodness of creation. Overemphasising it tends to eclipse the life of Jesus and can play down the positive aspects of creatureliness that are revealed in the incarnation.

Stewardship as the restoration of 'the order of creation'

The call to stewardship in creation theology refers to humanity's role as caretaker for all of creation. Northcott's view of stewardship can be reconstructed from his theological and ecological analysis; where nature is viewed as creation and is fundamentally ordered and relational. As a gift from God the land is destined for the care of the good steward until its eschatological restoration. The liberation of creation for Northcott consists in restoring the original priority of relationality. Eschatologically this implies a return to a paradise which is characterised by a 'natural relationality'. 'The *telos* of the cosmos is the restoration of paradise, of the natural relationality between humans and God' (193). In consequence

---

42 Northcott, 'Ecology and Christian Ethics', 214.

the role of the steward, it is implied, is to conform to this prior relational order, both divinely and naturally ordained.

The most significant characteristic of creation, in this analysis, is relationality which is interpreted as reflecting the being of God the creator and ordered goodness, revealed and vindicated in the incarnation and resurrection of Christ. In contrast to modern individualism and subjectivism, the 'Hebrew Bible offers a fundamentally interactive account of the relations between the human self, the social order and natural ecological order, and between all of these and the being of God' (164). There is, he argues, a 'deep ecology of the created order' (195). Scripture reveals that there 'is a critical order in the biosphere and humans have a fundamental duty to preserve this critical order, by abstaining from idolatry, and by tending the earth justly and with respect' (196). The natural and social order are therefore both aspects of creation with which humans should co-operate to maintain the relational balance.[43] The 'moral and religious aspects of human life tend to the same end, which is to preserve and restore the stability, harmony and relationality of all things' (197).[44] The 'cosmos' reflects and reacts to disturbances in this relationality and the 'disruptions of natural disasters are signs of disturbances in human and divine relationships, of human mismanagement of nature and society, rather than evidence of natural evil' (197). Humans must resist violence and domination in order 'to reconcile those things that are opposed'(198).

It could be said that this revealed, ordered, deep relationality, which is fundamental to his environmental ethic, owes more to Northcott's particular analysis of ecological concerns, and to his critique of modern individualism, than to systematic theology. His view of relationality and

---

43 Celia Deane-Drummond, however, points out that the ecological sciences have moved on from the 'equilibrium' view of natural systems, with its associated concepts of stability, harmony and balance to a non-equilibrium view that recognises that this harmony only applies under certain conditions and that natural systems also break down under internal and external influences. Cf. Celia Deane-Drummond, *The Ethics of Nature* (Oxford: Blackwell, 2004), 36–38.

44 He also suggests that a future peaceable kingdom may even transcend the original peace of Eden but his theme is predominately one of restoration (197).

hoped-for environmental harmony also founds his theological critique of individualism.

> Modern individualism arises ... from the distancing of self-consciousness from embodiment, and from the disembedding of the self from communities of place as traditionally constituted by the worship of God, and the correlative recognition of divine order in the cosmos and of divine intentionality in human society.[45]

Individual fulfilment is predicated, and reciprocally related, for Northcott, on a holistic harmony (269). Stewardship, from this perspective, is in the service of this harmony, holism and relationality. Yet, it is questionable that harmony, as we have already seen, is automatically the highest good for a Christian or an environmental ethics.

An evangelical reading of natural law

In his approach Northcott seeks to integrate Roman Catholic natural law perspectives with sociological, ecological and economic insights in the development of an environmental theology. On the one hand he appreciates the universal aspects of natural law approaches and the positive contribution of human reason and morality. However, he tends to naturalise these in his approach and appears to conflate naturalness (interpreted ecologically) with creatureliness. The 'moral order of the world', for Northcott, reaches its fullest expression in the work of Aquinas on the natural law. Northcott interprets it as that which reflects the biblical account of the fundamental relationality between the human person, 'the natural order', and God. The natural order, in turn, is that which is characterised 'by purposive order and equilibrium'. This natural order, he says, should exhibit relationality and equilibrium and is fundamentally linked with the social order, which it follows, and which should exhibit the same tendencies.

---

45 Northcott, *The Environment and Christian Ethics*, 205. Further page numbers are in the text.

For Northcott it is 'the natural law' which links the Hebraic images of primal order to the Christian belief in the redemption that is already begun in Jesus (226). This order in nature is recognised, for Northcott, by diverse cultures and can be found among contemporary environmental philosophers (165). He finds evidence for a natural peacefulness in other species and for environmentally harmonious morality in non-western tribes and human groups (176–187). The specifically 'Christian doctrine of natural law represents a belief in the moral purposiveness and relationality of the cosmos, and in the relation between the human quest for the common good and the goodness of the created order and the other orders of being which inhabit the creation' (164).

It is not always clear, however, what Northcott means by 'natural law'. It is used in some instances to refer to the regularity in physical reality, and at other times it stands as a model for ideal and harmonious living. This double perspective is a feature of the natural law tradition as we have already seen in chapter two. In Roman Catholic moral theology natural law can refer to the ideal rational nature of a human person or to his empirical structure as a natural being with needs, as Schockenhoff points out. Northcott appears to use these two meanings interchangeably. He says that in 'an ecologically revised Thomist perspective, we may say then that natural law is operative at every level of reality, and ecosystems like animals pursue certain goods after which they are teleologically ordered by the creator'.[46] Yet, he interprets this to mean that 'the world has an intelligible order to which it is our duty as humans to conform'.[47] This interpretation tends to naturalise natural law and to conflate physicality with human knowledge and morality, something that is expressly resisted by revisionist natural law ethicists, where 'nature' in 'natural law' is understood as 'reason'. Knowledge and morality are what distinguishes the human person from the rest of the creation in natural law, even as 'nature' implies, at the same time, that we remain embodied persons in continuity with it.

46 Ibid., 253.
47 Ibid., 254.

The natural law concept of the common good entails a critique of both individualism and globalisation which enables Northcott to identify a predominantly 'parochial ecology' of local worshipping communities as the most appropriate Christian response to environmental problems. He elevates the ecological insight of the interconnectedness and interdependence of all life to a social principle, giving priority to the role of intersubjectivity over self-reflexivity, the other pole in anthropology that also needs to be retained. For Northcott creaturely existence is revealed as properly interdependent rather than also independent.

Northcott rightly values the positive understanding of natural law as an ethic that insists on a moral order that is prior to human making and that is accessible to all. Yet, although he claims to adopt a natural law ethic, he immediately revises it ostensibly along ecological lines. 'The natural law tradition needs ecological revision, and cannot be applied in a fundamentalist way from medieval philosophy to modern ethical problems' (267). He brings together aspects of natural law approaches and ecology to develop his theologically informed environmental policies. He understands the common good as predicated on trust in a providential God who freely confers his gifts on all humankind. From Isaiah he draws on a vision of natural self-sufficiency and from the gospels the emphasis on the abundance of nature's gifts in Jesus' ministry. All of these serve to elaborate his central thesis, that stewardship is excessively anthropocentric and optimistic about human creativity. In contrast it is Sabbath rest, rather than human activity, that truly shows the human person's gratitude for the gift of creation.

In his reconstruction it is telling that although he refers to the concept of the common good and its corollary subsidiarity, he fails to draw out their implications in his 'parochial ethic' (237). He seems unaware of the possibility that an emphasis on 'holism', community and relationality is in danger of slipping into a type of 'Christian utilitarianism'; where, for example, it pitches the good of the individual against that of the community.

In addition, his passing reference to subsidiarity yields a common misinterpretation. 'Subsidiarity', he correctly begins

... argues for the legitimacy of local and natural associations: the powers of the state must always be held in check by the associations of workers, peasants and local communities through which the natural right of the person to livelihood and the necessities of life may be asserted and maintained against corporatism or statism. (237)

However, he concludes that this implies a 'parochial ecology' that is both socially and ecologically local and theocentric.

> The pursuit of parochial ecology will involve local Christian communities in a quest for just and hence alternative ethical approaches to global economic exchange *and* in efforts to promote the flourishing of the local human and ecological communities in which worshipping communities are situated.[48]

In turn worshipping communities must engage in a liturgical reform that attends to the cycles of the earth (321).

> The ecological reform of local and parish worship will involve the recovery of holistic rituals which reconnect worshippers to their kinship as embodied beings with the whole of creation and with animals, to the dependence of human society on the ecosystems which give life to humans and animals alike and the natural goods and goals by which all created life is marked by the purposive hand of the beneficent creator and is directed towards final restoration. (322)

Yet, as we have seen in chapter two, the principle of subsidiarity does not have to pitch the local against the global. As Alkire pointed out in regard to subsidiarity, 'the essential insight of this principle is that maximal responsibility for temporal affairs should be assumed by the most local organisations capable of carrying them out.'[49] The most 'local' institution could well be an international one and in some issues in environmental ethics it is likely that the most local institution is a global one. The

---

48   Northcott, 'Ecology and Christian Ethics', 225.
49   Debt relief cannot be solved by individuals, civil society or individual states. According to the principle of subsidiarity it 'bubbles up' to international institutions. Cf. Alkire, 'When Responsibilities Conflict: A Natural Law Analysis of Debt Forgiveness, Poverty Reduction, and Economic Stability', 73.

consequences of a requirement that international and global institutions be theocentric would raise many issues, not least for interreligious dialogue to address environmental issues.

In terms of environmental policy, Northcott proposes the 'democratic regulation and control over markets, technologies and resource use' (310). In economics he critiques wealth acquisition for its own sake (283) and expresses a preference for the stability of a 'steady-state' economy over 'growth' economies (286). He adopts a natural law approach because he says, 'Thomism' has generally been subversive of dehumanising regimes. It has

> ... been generative of political ideals and ... a consistent source of criticism of the excesses and human indignities perpetrated under various modern political and economic practices and ideologies, including industrial capitalism, laissez-faire economics, communism and fascism. (237)

This disguises the historical complexities of the impact of Roman Catholic natural law ethics on society and politics.[50] In several aspects of his environmental ethics he uncritically marries theological insights with aspects of ecological thinking that are questionable, theologically, philosophically and scientifically.

Conclusions

Northcott's approach begins with the systematic theological point that there is order in creation that is divinely ordained. He translates that, in his environmental ethics, into a call to conserve the prior order in nature and creation. This order, he suggests, was disrupted in the degeneration of the early creation–redemption synthesis of Irenaeus, through subsequent Romanisation and Hellenisation. Yet, as we have seen from Pannenberg's analysis, the Christian doctrine of creation was developed in a critical overturning of implicit philosophical assumptions in Greek

---

50 Cf. MacNamara, *Faith and Ethics: Recent Roman Catholicism.*

thinking by the biblical understanding of the being of God and the nature of God's creation. The development of an environmental theology, by critically using current philosophical thought forms, is a continuation of that process.

Northcott borrows from O'Donovan's evangelical ethics when he emphasises the resurrection as the moment in Christian history, admittedly alongside others, that vindicates the goodness of creation. The resurrection serves as the Christological principle that allows him to reintegrate an evangelical and a natural ethic. However, he confuses the goodness and givenness of the created order (unity) with natural order. This is problematic for both theology and ethics. The incarnation and resurrection are a revelation of God's essence as love and his power of renewal even in death. Northcott diffuses their significance for Christian creation theology by privileging the restoration of created order over its transformation.

From an ecological perspective Northcott's interpretation of the natural order is too closely tied to revisable concepts in the natural sciences, such as the concept of order and equilibrium in ecology and steady-state markets in economics. Celia Deane-Drummond for example suggests that 'the cultural belief in the balance of nature' is an 'ecological myth'. Contemporary ecology also recognises that systems may not exhibit order and balance, that they can fail catastrophically, even in the absence of human intervention.[51] She says 'to pose the organic model of the earth in opposition to the mechanistic one only takes us to a new set of problems. Both models suggest a fixity of order that is only part of reality'.[52] Ecological reconstructions of 'nature' cannot be simply tagged onto theological interpretations of creation in environmental theologies.

For Northcott, the failure of anthropocentric instrumentalism and reductionism leads him immediately to praise holism and relationality. He considers 'holism' to be morally superior to other perspectives. Yet, as Pannenberg suggests, interdependence is like independence; they are

---

51 Deane-Drummond, *The Ethics of Nature*, 39.
52 Celia Deane-Drummond, *Creation through Wisdom* (Edinburgh: T. & T. Clark, 2000), 246.

both aspects of creatureliness. Both are expressions of our finitude, neither of them are evil but they both bring conflict which could be the source of failure and sin. Northcott is correct then to say that interdependence is constitutive of creaturely existence, just as environmentalists would emphasise. However, rather than being morally superior, our constitutive interdependence carries with it theological ambiguities in relation to fallenness that applies to all of creation; in its interdependence as well as its independence. Northcott claims that his

> ... approach to ethics contrasts dramatically with the modern secular view of human ethics as a project of creative construction which is undertaken in opposition to the prior order which humans encounter in the physical nature of things.[53]

However, in his attempt to show secularisation as a degeneration of relationality and community and of the early Christian creation–redemption synthesis, he fails to integrate positive interpretations of human creativity that are present as contrasting but abiding strands in Christian creation theology itself.

Nature as creation in the 'virtue' ethic of Celia Deane-Drummond

The critique of technology and of instrumentalism is rightly a common thread in Christian environmental theologies and is a theme that is shared in the work of Celia Deane-Drummond. In this section I will present those additional aspects of Deane-Drummond's work that might contribute more specifically to the question of transgenic technology in food production. Her critique of the interventionist attitude in human relationships with 'nature' is mounted from the perspective of a virtue ethics which integrates insights from philosophy and the natural and

---

53   Northcott, 'Ecology and Christian Ethics', 216.

social sciences as well as from theology. She links her virtue approach in ethics, to a theo-ethical interpretation of Wisdom; interpreting Wisdom as a divine revelation of the prior goodness and giftedness of creation but also as the principle of creative transformation.[54] For her, practical reason – as a character of the rational human agent – has a theological analogue in the concept of 'practical wisdom'.

In this section I will examine how she reads the conflicts over new genetic theologies as an opportunity for theology, in the sense that theological ethics has the integrative tools to conceptualise these conflicts. In keeping with many theological responses to environmental problems she highlights the wisdom tradition as the most adequate perspective from which to critique instrumentalism in her reconstruction of stewardship. She argues in favour, on the basis of the wisdom strand to be found in systematic theology, of the human creative capacity for virtue and for the good life in ethics. This is a theme also shared, although with qualifications, by the second generation autonomy position in Christian ethics.

Theology as an interpretative resource in conflicts over genetic technologies

Celia Deane-Drummond et al. suggest that the disciplines of theology and ethics can help to conceptualise and interpret public 'disquiet' over new genetic technologies, in particular in food production, in the public realm. Genetic technologies, they suggest 'involve profound challenges to our ways of thinking about human identity and our relationship with the natural world – challenges that have to be seen as at least implicitly theological in nature'.[55] The identification of aspects of the debate with theology arises in part because public concerns about genetic technologies are often expressed, for Deane-Drummond et al., in theological language. The 'ambiguity in human attitudes to science manifests itself in religious traditions and discourse, and this is certainly true in the current mixed

---

54 Cf. Celia Deane-Drummond, *Genetics and Christian Ethics* (Cambridge: Cambridge University Press, 2006).
55 Deane-Drummond and Szerszynski, eds, *Re-Ordering Nature*, 2.

responses to the new genetic technologies.'[56] It is also the case that this disquiet over the genetic modification of nature is not only expressed in religious or quasi-religious language but, for Deane-Drummond et al., reflects 'deeper religious issues about the ordering of the natural world, the place of God in such an ordering, and the nature of human identity' (9). Given this analysis, they suggest that the 'deeper issue' at stake is an anthropological one; of how the 'new genetics' impacts on human self-understanding. Transgenic technology has prompted a debate about what it means to be human in relation to the non-human natural world:

> ... the intensity of current controversies around GM crops and foods arises in part from the fact that, in their regulation in the public domain, *conflicting ontologies of the person* are making themselves felt in the politics of everyday life. (22)

There appears to be an unbridged dichotomy between public (or popular) and institutional and corporate conceptions of the risks of GM food, both to human health and environmental wellbeing. The very institutions responsible for these technologies are not capable of recognising or conceptualising such a dichotomy.[57] From this sociological analysis, the authors confidently conclude – in contrast with Northcott's ecclesial approach, that retreats to the specific domain of the churches – that theology may be essential to the debate. Theological 'perspectives may now be indispensable in helping explain to largely secular institutions the sources and dynamics of conflicts now threatening to paralyse the development of what is being posited as a key technology for the twenty-first century' (22).

GM technology has furthered the shift towards the social recognition of uncertainty (314). Deane-Drummond et al. contend 'that this shift is both a problem and an opportunity for religion'. It is no longer possible to leave nature to science, society to politics, and religion to personal morality. What they propose is a relatively modest programme that sees an 'opportunity and challenge for religion in general and theology in

---

56   Ibid., 4.
57   Cf. Ibid., 40.

particular – to operate once more in a much wider cultural space than has been the case for the last few centuries' (314).

Christian creation theology can conceptualise the disjuncture between the experience of reality as fundamentally ordered but at the same time full of creative possibilities for human ingenuity. It is in this, a perhaps less direct sense than Deane-Drummond might allow, that creation theology, as a hermeneutical discipline, can contribute to the debate about GMOs in the public realm. It is clear from their sociological analysis that public responses to GM food challenge the sense of order that people experience in fundamental categories of 'nature'. Even though it can be tacitly assumed that creative interaction with the natural order can be justified in certain circumstances, the justification for GMO releases has been unconvincing so far, given the perceived dangers.[58] The concerns of ecology, sociology and theology converge on the question of the limit to instrumental reason, but each conceives of it in significantly different ways. I will examine in this next section her reconstruction of stewardship as a cultivation of the virtues.

Stewardship reconstructed as the cultivation of the virtues

Celia Deane-Drummond, like many environmental theologians, founds her view of nature as creation on aspects of the biblical Wisdom tradition.[59] She draws aspects of her critique from the theology of Jürgen Moltmann and from feminist theological perspectives. As a scientist she is also interested in the contribution that non-theological disciplines, such as sociology, can make to theology and to ethics. In this section I examine

---

58 Celia Deane-Drummond, Robin Grove-White, and Bronislaw Szerszynski, 'Genetically Modified Theology; the Religious Dimensions of Public Concerns About Agricultural Biotechnology', *Studies in Christian Ethics* 14, no. 2 (2001), 31.
59 Deane-Drummond, *The Ethics of Nature*, 219. This book draws on her earlier work in Deane-Drummond, *Creation through Wisdom*. Further page numbers are in the text.

her appropriation of the systematic theme of wisdom in developing her environmental theology and her theological reconstruction of stewardship as a cultivation of the virtue of wisdom.

An environmental theology based on wisdom

What is significant for this study is that Deane-Drummond interprets Wisdom not only as the prior order in nature, as we found in Michael Northcott's analysis, but also as God's continuing activity in and through creation. She links Wisdom's concern about the divinely posited natural order with the logos theology of the New Testament. We 'find wisdom taken up in the New Testament to refer to the action of God in Christ, through the wisdom of the cross, as well as more obliquely in the Logos language of John' (220). This allows her to recover a theological view of nature as creation which prioritises its givenness and emphasises human dependency on God's preserving will. However, it also leaves room for a positive evaluation of creaturely participation in God's preservation and governance of creation. The Wisdom literature clearly reflects the order in creation, the order that provides a link between humanity and all the rest of creation in Old Testament scriptures. However, it is a diverse body of literature and Deane-Drummond highlights the importance of wisdom as a motif where it leaves room for God's continuing participation in creation (221).

A wisdom-virtue ethic, she proposes, is the bridge between science and faith. 'Wisdom is ... the virtue that can express the mediation between scientific reasoning and faith, between the discoveries of the creaturely world and the telos of the created order as participation in God' (20). It is likewise the bridge between faith and ethics. The logos, Sophia, is 'particularly useful, in that it can act like a bridge between the philosophical concepts of prudence and theological ideas about who God is as Creator'(43). In this regard she says, that while 'prudence is the practical grounding for ethical decision making, wisdom is its theological counterpart, encouraging the search through fellowship with God' (44). For Deane-Drummond, wisdom is the bridge between the speculative and the practical. 'Hence wisdom is not just knowledge about God in a speculative sense, but is both speculative and practical, located in the context of salvation history and carrying certain moral obligations' (45). In this she is close to Pannenberg's understanding of God's free action in the creation, preservation and rule of the world and of redemption as a transformation of the created order rather than its mere restoration.

Deane-Drummond captures Pannenberg's systematic concern to keep the order in creation and the transformative power of the Christian God in tension. However, in her wisdom approach she also unites in confusion several disparate strands in theology under the one theme of wisdom. Wisdom is an attribute of God. It represents the feminine face of God (221). It represents God's relationality with the world (221). It is also a virtue that needs to be fostered in community. In that sense, she says, it challenges the individualistic way we think about ourselves (220). It can integrate insights from other disciplines in the human sciences and it has commonalties with sapiential understandings in other religious traditions (220). This confusion may be a reflection of the fact that the Wisdom texts represent distinct literary forms that provide different insights into God's relationship with the world in changing historical contexts.[60] However, the appropriation of the wisdom strand in systematic theology for environmental theology needs further clarification.

Stewardship and virtue ethics

For Deane-Drummond, as for Northcott, the biblical notion of stewardship needs ethical and theological revision if it is not to lapse into instrumentalism. Rather than contrasting stewardship with Sabbath rest and a return to an original order as Northcott does, Deane-Drummond instead builds her anthropology on the active participation of the human person in creation. Human participation in creation, in knowledge and morality, are interpreted through the lens of a virtue approach.

---

60 Clifford, 'Creation', 206.

Virtue ethics may appear, she says, at first glance, to be too anthropocentric to be helpful in developing an ethic of nature.[61] However, Deane-Drummond suggests that a virtue ethic allows us to extend the moral community to non-human reality; to develop a more communitarian and 'holistic' reading of nature, alongside other perspectives, so that creatures are valued alongside the human community (226) and we widen the circle of care (228).

Her search for a Christian account of the virtues is in keeping with her concern to develop a Christian anthropology that opens a window on a full account of human flourishing in continuity with the natural world. She is aware as a scientist, and particularly from her sociological sources, that technological changes have significant structural implications, that nature is also socially constructed. In relation to biotechnology, for example, she understands that changes to 'natural' goods affect the 'givenness' of social and institutional structures which in turn inform and constitute the possible moral choices of virtuous agents. Theological insights and critiques can contribute to making aspects of these reconstructions more explicit.

Stewardship is reconstructed implicitly in Deane-Drummond's environmental theology as both a gift and a task. This is one of the insights of her wisdom critique. Wisdom theology points her towards a concept of stewardship that involves a cultivation of both the philosophical and theological virtues.

A virtue ethic of nature as creation

Deane-Drummond's virtue ethics approach is built directly on her analysis of the Wisdom strand in biblical theology. She clearly differentiates between Christian and non-Christian interpretations of the virtues but she suggests that these inform each other rather than being in conflict

---

61 Deane-Drummond, *The Ethics of Nature*, 18.

with each other. She, like Northcott, borrows from Aquinas, but from Aquinas's analysis of the virtues rather than specifically his natural law.

The basic premise of a virtue ethic, it is assumed, is that goodness is a more fundamental consideration, rather than rights, duties or obligations. A virtue ethic focuses on the agent rather than the action, the character of the person rather than the rule to be followed.[62] Deane-Drummond draws on the four philosophical virtues of Aristotle, namely prudence, justice, fortitude and temperance, in so far as they were later developed as cardinal virtues by Aquinas, who added the theological virtues of love, faith and hope. She suggests these are not in conflict with each other but are mutually enriching. She gives priority to prudence over the other cardinal virtues and to love in the theological virtues. According to her interpretation of Aquinas the 'form of prudence to which the Christian aspires is prudence in the light of the three theological virtue of faith, hope and charity [love]'. This is in keeping with the understanding in Christian ethics that love relativises all other ethical positions. She echoes this when she says that the ultimate goal for a Christian is not goodness in an abstract sense, but friendship and participation in the life of the Trinitarian God (13).

Although Deane-Drummond does not develop any account of fallenness and sin, she does imply that a theological interpretation of the virtues is needed in order to deal with 'the brokenness that is the inevitable part of searching to gain a virtue' (7). In addition a properly Christian discussion of the classical virtues would include virtues that are specific to a Christian understanding of being created; the need for forgiveness, for reconciliation and for God's grace (8).[63] A theological interpretation of the virtues also has to investigate its biblical foundations and it is to this aspect of Deane-Drummond's analysis that I now turn.

---

62 Ibid., 6.
63 Deane-Drummond acknowledges that 'the extent of evolutionary "waste" and the raw suffering in nature remains a challenge to those who affirm belief in the goodness of God and creation' (224). She offers the 'wisdom of the cross' in answer to the ambiguity of suffering, without further development (224).

A biblical basis for a Christian virtue ethic

Although Deane-Drummond reconstructs her virtue approach along specifically theological lines, she has to contend with the challenge from biblical scholars that contemporary virtue approaches do not appear to have clear biblical foundations. This is a significant issue for any theological ethical position that wants to develop an ethics of the good life and of virtue.

John Barton questions the assumption that a virtue ethic, as understood in contemporary ethics at least, is clearly reflected in the biblical texts.[64] He suggests that the Old Testament in particular presents us with models for conversion rather than with the sort of models for growth in moral life that can be found in virtue approaches. As Deane-Drummond notes from Barton's analysis:

> The Old Testament presents us with the stark contrast of the wise or foolish, the righteous or sinners. Characters seem to be fixed and unchanging, ethical choice seems to be once and for all, where 'the subtlety which sees everyone as a mix of the two, or as living a life in which virtue is cultivated and vice therefore rooted out seems largely lacking'.[65]

The emphasis, in this context, is on conversion rather than growth in moral life; it is deontological and legal in character.

However, Deane-Drummond observes that the wisdom literature also provides alternative models to conversion. 'Qoheleth and Job are both excellent examples of those who challenged the presumed wisdom of their contemporaries, arriving at more open-ended accounts of the way God works in the world'.[66] In the NT the 'idea that external action

---

64 John Barton, 'Virtue in the Bible', *Studies in Christian Ethics* 12, no. 1 (1999).
65 Celia Deane-Drummond, ed., *Aquinas, Wisdom Ethics and the New Genetics, Re-Ordering Nature; Theology, Society and the New Genetics* (Edinburgh: T. & T. Clark, 2003), 295.
66 Ibid., 296.

flows from internal dispositions of character is a thread running through Matthew's gospel; for example, Matt. 3.8, 10; 7.15–20, 12.33'.[67]

Barton also admits that although there may not be an explicit virtue ethics in the biblical narrative, there is an 'implicit' one in testimonies to the experience of self-transcendence. The biblical narratives of David, for example, show a human person, not an exemplar of an abstract virtue, but as a concrete individual with perhaps a deeply flawed but nonetheless examined life. This, he suggests, illustrates the difficulties of the moral life, the difficulty of contextualising our moral duty, given our finitude, frailty and failure.[68] It is this understanding of the virtues that is of interest to Deane-Drummond. It is one that also resonates for a Christian reconstruction of autonomy that understands that duty and obligation needs to be both contextualised and reinforced at the level of values.

Summary and critique

Celia Deane-Drummond is concerned theologically to develop a sapiential ethic of 'discerning wisdom' capable of sustaining a critique of instrumentalism in relation to 'nature', both human and non-human. This allows her to retain a respect for the insights revealed by the natural sciences and indeed to give due praise to those aspects of biotechnology that are ethically uncontroversial. Her virtue approach integrates the theological and philosophical virtues in mutual critique.

As a plant physiologist and theologian, she is open to scientific understandings of 'nature' as intellectual resources for Christian theology. Environmental problems, she suggests, challenge all prevailing instrumentalist anthropologies that neglect human continuities with non-human reality. Theologically she is concerned to relate human creativity in the sciences with the role of the human person in the creative work of God.

---

67   Deane-Drummond, *The Ethics of Nature*, 9.
68   Barton, 'Virtue in the Bible', 19–20.

In her response to the challenge of the new genetic technologies, she reconstructs her anthropology from a virtue ethics perspective. For Deane-Drummond a virtue approach in ethics contributes to an ethics of nature by contributing to what it means to be human. It is a teleological approach, since it elaborates on human conceptions of the good life and accounts of human flourishing (2).

Deane-Drummond, however, does not simply assume that virtue approaches in ethics are automatically more Christian that other approaches. They also need to be evaluated from the perspective of Christian ethics. Christian ethics, she suggests, has alternative notions of virtue that are subversive of classical Aristotelian conceptions of virtue which are based on personal merit and social status. Virtue theory needs to be critically appropriated in ethics and integrated with specifically Christian components such as forgiveness and reconciliation. Referring to Aquinas's analysis of the cardinal virtues as her touchstone, she suggests 'that the combination of stillness and action, of appropriate reflection and concerted judgement, find their most cogent expression in the theology of Aquinas, in particular his understanding of the virtues and practical wisdom or prudence.' (ix)

In relation to biotechnology, she analyses the way in which new technologies impact in the public domain and how they impact on human self-understanding in relation to the natural world. Theology can contribute to the ethical discussion around new technologies in its critique of instrumentalist anthropologies and conceptions of non-human reality. In addition, Deane-Drummond implies, the virtuous moral agents can reorient themselves towards environmentally sustainable living and in turn, environmentally sustainable living is transmitted through value communities.

Pannenberg argued that wisdom theology is concerned with the question of creation's order. On the one hand it seems therefore, that all that the discerning virtuous agent can do is to be concerned with perceiving this order and living in harmony with it. This is the approach that, as we have seen, Michael Northcott takes in his environmental theology. However, the Wisdom texts also represent distinct literary forms that provide different insights into God's relationship with the world. It is not, therefore, immediately obvious that these alternative world views can be simply harmonised and integrated.

However, in developing a wisdom ethic Deane-Drummond seeks to keep in tension the ambiguities in creation theology that point to both the affirmation of the original goodness and givenness of the created order and the Christian insight that redemption implies more than mere restoration of dominion. She is closer to Pannenberg than Northcott is in her understanding of the recreative as well as receptive capacity of the human person in God's continuing creation.

In common with the autonomy approach, she points to the relativising dimension of a Christian faith context.

> ... it is the overwhelming love of God that relativizes all human efforts, while at the same time holding a clear perspective on human obligation and responsibility. I prefer to call this form of prudence wisdom, since it can be more readily identified as a theological category. While prudence is in one sense a form of wisdom, wisdom in the end moves beyond even the scope of prudence.[69]

From the perspective of an autonomy position, the strengths of her approach are that she develops the link between the personal virtues of the individual and the structural possibilities and limitations that living in a particular community, professional expert culture, private enterprise or state institution might engender. However, although she points to a transcommunitarian account of the virtues in her discussion of justice (15) she does not convincingly get beyond a context-bounded definition of the virtues to the level of moral obligation. She eschews Onora O'Neill's proposal to reintegrate justice and virtue in an account of moral reasoning and proposes instead that justice be considered as a virtue. From an autonomy perspective her emphasis on the 'practical wisdom' of the virtuous agent, cannot be comprehensive of the issues at stake. It is not just a question of how the virtuous agent or indeed virtuous scientist would act, but how virtues are sustained and interpreted, in continuity and change as well as transcended towards a different level. From an

---

69   Deane-Drummond, *The Ethics of Nature*, 14.

autonomy perspective in Christian ethics a virtue approach needs to be integrated into a deontological framework.

Towards a Christian anthropology of stewardship

Christian creation theology reconstructs nature, as it is understood in the natural sciences, as the good creation of a benevolent God destined for salvation and fellowship. In dealing with creation as the free act of God, Pannenberg distinguishes between the inward and outward action of God, the nature of creation, its Trinitarian origin, and the related concepts of God's preservation and rule of the world. These imply that the Christian God created not out of neediness but out of love, and that creation therefore is originally good. They also imply that creation is destined for redemption in divine fellowship and not merely for restoration to an original state. In relation to the time-bound world of creatures, Pannenberg suggests that creatureliness implies both order and unity, as well as contingency and plurality. This has implications for environmental theologies, in that, as part of our finitude, both independence and interdependence are aspects of a Christian ontology of creatureliness. Our creatureliness, as independent and interdependent creatures in the plurality of creation, is not sin but neither is it yet fellowship with God. However, our human freedom and our relationality are abiding aspects of our creatureliness. Both of these influence Christian reconstructions of liberation for the human person and for the nonhuman creation in environmental theology.

Both Northcott, in his 'natural law' ethic, and Deane-Drummond in her 'virtue' approach begin with a critique of the negative aspects of anthropocentrism and Christian betrayals of the call to stewardship. Yet, they conclude with quite different proposals in terms of environmental policy. Northcott emphasises the failure of humanity, in modern agriculture and technology, to live up to the call to rightful stewardship. He reconstructs an evangelically interpreted non-fundamentalist 'natural law' perspective in the interests of re-establishing a moral significance for non-human reality in Christian ethics. It seems that for him the liberation of non-human creation can only be effected theocentrically through the reinvigoration of 'local worshipping communities' who will predominantly practice 'parochial ecology' in the face of global environmental decay.

Deane-Drummond, in distinction, analyses the way in which new technologies impact on human self-understanding and on how society might engender the conditions under which virtuous moral agents can reorient themselves towards the goal of environmentally sustainable living. To this end, she develops a 'virtue' ethic of the environment that is rooted in the OT wisdom strand of biblical environmental theology, but interpreted through the incarnation of Christ as the Logos of God. Her wisdom ethic implies a more positive evaluation of human knowledge in modern science and technology – where it is allied to wisdom and 'practical reason' – than found in Northcott's account. In terms of practical policy, her virtue approach points to the need for institutional change that can foster the virtues that contribute to sustainability and development.

These approaches deliver lessons for the reconstruction of an environmental theology from an autonomy perspective in Christian ethics. Both Pannenberg's systematic analysis and the critique of environmental theologies from some of his insights suggest that, from an ethical perspective, freedom and autonomy capture the insight of Christian theology and of creatureliness more adequately than either the categories of nature or disposition that are found in Northcott's particularist exposition of natural law or Deane-Drummond's wisdom-virtue ethic.

Christian models of stewardship have the distinct task of continuing to integrate the free creaturely participation in the creative preservation and rule of a good and caring God, who is Himself, 'free from envy', with a recognition of our fundamental frailty and dependence. The ongoing task for environmental theology is to continue to elaborate on what being made in the image of God implies for the concept of stewardship. Stewardship is first a gift but also a task. Pannenberg rightly interprets Christian engagements with philosophy and science not as automatically compromising biblical distinctive content but rather as a complex, often positive, historical process. He has had a sustained engagement with the natural sciences in his theology.[70] For Pannenberg theology cannot 'refrain from describing the world of nature and human history as the creation of God'.[71] Mostert, in describing Pannenberg's theology as 'an expanded doctrine of God', suggests that Pannenberg sees creation as an open system, open to a new, including a higher, consummation.[72]

What is significant from his analysis is that the enduring basic relationship of humanity to God is one of creatureliness. It is important to elaborate what this implies for Christian stewardship. Langemeyer suggests that in contemporary Christian anthropology, creatureliness can be characterised in many relevant ways. It can be described as

> ... the accidental and arbitrary character of existence (contingency); as the temporally scattered, split-up nature of existence (finiteness); in the sense of the basic human neediness (for food, shelter, social contact); in the awareness of the historically situated limitedness of the possibilities of existence and its realization ('to be lucky or unlucky'); as insecurity (anxiety) and the experience of being easily led astray.[73]

The first creation account holds the human person up as the crown of creation, the best of all God's works. The second describes creatureliness as transitory, needy, susceptible to pride and hubris.

> Theology has found it difficult to bring together both aspects of the saving work of Christ: on the one hand, the affirmation (restoration) of creatureliness ... and, on the other, the elevation of creation – which was ruined by original sin – beyond itself to participation in God's eternal life.[74]

Yet, this is the task for creation theology.

---

70 Clifford, 'Creation', 238.
71 Pannenberg, *Systematic Theology Volume II*, 59.
72 Mostert, *God and the Future*, 175.
73 Georg Langemeyer, 'Creatureliness of the Human Person', in *Handbook of Catholic Theology*, ed. Wolfgang Beinert and Francis Schüssler Fiorenza (New York: Crossroads, 1995), 153.
74 Ibid., 152.

When the fallenness of creation is stressed it is the 'figure of Prometheus, who epitomizes the very notion of self-creation, that is often played off against the creator-God of the Christian creed'.[75] However, as Oswald Bayer points out, this is unjust because Prometheus 'revolts against a jealous and merciless Zeus and not against the One who – as the Almighty and at the same time Merciful One – grants the *dominium terrae* to humankind, entrusting them with it, all the while remaining himself free from envy'.[76] How do we balance creaturely participation in the creative preservation and rule of God with a recognition of our fundamental frailty and dependence? For Christian anthropology the mistake has been to overestimate our prowess. However, it is also a mistake to miss the potential in the human capacity to be both receptive and constructive. Undoubtedly 'a soteriological over-estimation of technology can only be sharply rejected; nevertheless, the inventive and constructive aspects of human dignity to reign over itself and over the world must also be given their due'.[77]

A Christian construction of stewardship has to do justice to the insights of biblical theology and to the coherence of systematic theology before it is applied in ethics. It is not reason as such, nor intellectual ability but the destination for fellowship with God that theologically gives humanity dominion in the time-bound world of creatures. The wisdom tradition draws our attention to the prior awe, beauty and order in nature that is not of human making. It is not, however, a replacement for the stewardship strand in systematic theology. The human person who is destined for fellowship with the Christian God 'who makes everything new' remains responsible for the liberation of all creation. In the context of current environmental problems pointing to the prior order and goodness of creation is a worthy corrective for instrumentalist approaches to nature. However, creation theology points not only to the preservation of the order in nature but also to the possibility of liberation for all creation in advance of a hoped-for salvation.

---

75   Oswald Bayer, 'Self-Creation? On the Dignity of Human Beings', *Modern Theology* 20, no. 2 (2004), 275.
76   Ibid.
77   Ibid., 276.

Conclusion

The autonomy position in ethics argues that the category of freedom has great potential to capture theological, philosophical and scientific insights into what it means to be a finite created being. In its theological reformulation, it contributes in specifically five ways to moral and ethical evaluations. First it intensifies, secondly it motivates and thirdly it provides heuristic potential in the debate about the new genetic technologies. Fourthly, it has the capacity to integrate insights from diverse interdisciplinary sources, and to integrate ethical insights at different levels, those of obligation and of value. Finally, it has a relativising capacity which subordinates obligation to a theological view of the human person that sees the need for redemption as more foundational.

Firstly, the Christian context provides frameworks which keep open the possibility of human transcendence of personal finitude and of transcendental conceptions of human existence. Contemporary environmental ethics, for example attempts to capture the value of non-human natural reality that we encounter and experience as a given, rather than as something we construct and invent. Christian creation theology prioritises the givenness of created reality and of our creatureliness. For Oliver O'Donovan this givenness can be understood as exactly that aspect of reality that is not dull or predictable but which is spontaneous and not put into reality by us. The uniqueness of each created individual in passing on life and contributing to the unity of creation, for example, prompts an appreciation for all natural life beyond its value for human needs or desires. In this sense we can say that the Christian context intensifies our sensibility to non-human nature in environmental ethics.

Theological ethics is also sensitive to contemporary conceptualisations of the dignity of the human person and the manner in which that is reconstructed in relation to current ethical issues. The autonomy approach in Christian ethics interprets the secular realm as a theological resource that challenges theology to continue to investigate its own findings and presuppositions, and renew its own responses in the critical appropriation of those resources. Sensitised to the question of the full dignity and freedom of the person, theological ethics can, for example, appreciate sustainability as a partial scientific conceptualisation of the space in which human dignity develops, in the environmental domain, the creation of just institutions as an aspect of that dignity in the sociological domain, and development-as-freedom as an aspect of that dignity in the economic domain. This transposition to contemporary categories intensifies rather than detracts from Christian sensibilities since it articulates them in relevant ways. Yet, it is because of a theological appraisal of human dignity, following from being made in the image of God, that De Schrijver is keen to investigate the evidence and possibilities open to us and available in non-theological intellectual resources.

Secondly, the autonomy perspective has motivational potential because it demands that we both translate and critically re-appropriate contemporary interdisciplinary and ethical categories for Christian ethics. More importantly, however, autonomy has the potential to motivate our actions and dispositions, despite our obvious frailty and failure, in solidarity with past and future generations. Kant's 'antinomy of practical reason' points to the costs of our unconditional convictions, the realisation that the fulfilment of our obligations may not coincide with our happiness. This implies that an autonomous ethics is limited by success calculation, as Junker-Kenny points out. It can anticipate the greater possibilities of God who completes the intentions of our actions. It is particularly in cases of conflict, that the significance and costs of being willing to forego the advantages and promised improvements in technology can be grasped. Generating demands for all the possible advantages of promised technologies – despite their being destructive of finite resources – may be incompatible with other ethical goals, such as that of development and distributive justice. A readiness to maintain costly convictions can be supported at the level of values in Christian communities. Christian concepts of the flourishing life allow us to abstract from a personal experience of failure (or success) in solidarity.

Thirdly, the autonomy position in Christian ethics also has heuristic potential in that the Christian faith alerts us to what is morally relevant. It makes us aware that historical perspectives are often formulated in response to questions and world views that were current in their time and can only be evaluated in that context. It also points to the achievements and the mistakes of past paradigms that can enlighten current perspectives. Twentieth century environmental theology, for example, mounts a critique of the instrumental destruction of natural resources from the perspective of the divinely given order in creation, attested to in the biblical Wisdom literature. This has the important function of resisting the current danger of the instrumentalisation of 'nature', both human and non-human.

An exclusive focus on the order in creation, however, also brings attendant problems for systematic theology. It can restrict God's creative will to the mere restoration of creation rather than pointing towards the possibility of fulfilment in its transformation. Undoubtedly it highlights the receptive aspect of human participation in created reality but it tends to obscure the reconstructive aspects. In Pannenberg's creation theology we can see that there is an abiding tension between the concept of creation as divinely and originally ordered and of creation as also contingent and dependent on the continuing creativity of a faithful and loving creator. The Christian idea of a creating and redeeming God implies a notion of fulfilment that is more than just the naturalisation of redemption in a utopian restoration of the original order. The human person, made in the image of God, participates in God's creative transformation of reality in knowledge and in morality. God's creative participation in creation is a condition for the possibility of human freedom, rather than being set in opposition to it.

We cannot ignore the constructive aspects of human creativity, even if there is always a danger that they will be misappropriated. Rather than relying exclusively on the model of Sabbath rest, environmental theology still has the task of conceptualising a non-instrumental view of stewardship. The concepts of the public good in environmental ethics and the common good in theological ethics have mediating potential in that regard. They underscore the contribution of institutions and communities that engender values and visions of the good life and that are more than just a reflection of prevailing hegemonic values. Institutional ethics, in the public realm, which are highlighted in this study by both Mieth and to an extent by Deane-Drummond, reveal some of the gaps in decision-making and policy formation in relation to environmental questions. Institutions both reflect and construct values. They need, therefore, to go beyond 'aggregating preferences' to consider value frameworks, in order to continue to meet the needs and to reflect the values of communities and grass-roots movements which are essential to consciousness-raising. In the face of the challenge of environmental degradation, theology has a non-instrumental concept of nature as creation, and of stewardship as both a gift and a task. It resists surrendering the aspect of gift, or of task, to either a theological conservatism or to an instrumental reductionism.

Fourthly, the autonomy approach has integrative potential for ethics. It argues for a rapprochement of a deontological framework alongside teleological aspects of ethical reflection. Deontological language alerts us to the duty to which we are called, morally and theologically, against what we might have otherwise assumed to be obvious. Teleological language, in contrast, stresses the connection, not the disjuncture, between what we experience and what we validate and re-imagine, in realising our moral obligations. The Christian autonomy approach presupposes that faith and reason are compatible and are not simply always in conflict. Theologically this implies that while we need conversion, we also need ongoing ethical formation and the support of value communities. In this way we can say that the level of values supports and reinforces the level of obligation and vice versa.

Theological categories in autonomous ethics have integrative capacity in that they drive the discussion beyond the compartmentalisation of perspectives, which although necessary in themselves, are nonetheless partial. The conductive method in autonomous ethics, outlined by Mieth, has the capacity to integrate and prioritise insights across several disciplines. In relation to transgenic food it is not necessary to begin from a position of unconditional endorsement or of unconditional rejection. Rather, a conductive approach to plant transgenics analyses specific concrete relative strands and builds a cord of arguments that would justify the acceptance or rejection of this technology. The current evidence from both breeders and economists establishes that transgenic technology has no current potential to feed the hungry in the developing world, or to contribute to a more sustainable agriculture in the developed world. Although the arguments that have been put forward here may not individually dismantle every counter-argument, they are sufficient to reject this technology at present in the context of Irish agriculture, and in the context of the appropriate transfer of technology to the developing world. The scientific evidence and ethical analysis does not support the conviction implicit in the recent policy document on the coexistence of GM and non-GM crops in Ireland, that politics should commit present and future generations so uncritically to transgenic agriculture.

Finally, the autonomy approach in Christian ethics relativises morality. It reaches beyond both normative and contextual perspectives to the question of transcendence and meaning. Human self-transcendence draws us beyond attainable or unattainable goals to an expanding horizon. This is conceptualised in transcendental theology as an exploration of those conditions that make liberation possible but also point beyond liberation to a greater fulfilment and promised salvation. From that perspective we can affirm that nature is the true context of human life, and that this is reflected in Jesus' understanding of his own creatureliness, in his ministry and in his appropriation of his own inherited scriptures. Theologically we are called to be the good stewards of creation and to work towards sustainable development in environmental ethics. From an autonomy perspective it is possible to see the full cost of a pared-down instrumentalist interpretation of human creativity and of natural reality, that betrays our creatureliness and the call to be stewards of all creation. It is also possible to evaluate and appropriate contemporary intellectual sources for Christian ethics, so that the Christian context can intensify, motivate, provide heuristic potential for, integrate, and relativise our moral and ethical positions.

The Christian understanding of creation and creatureliness has profound implications for environmental theology and ethics. It changes the context in which environmental concerns are evaluated and legitimated. Christian stewardship should not be construed as endless service to 'nature', it is not a type of environmental protectionism. Nor does it suggest that individuals and their communities can be sacrificed to environmental goals. It does not envisage the human person as antithetical to 'nature'. Indeed it is open to the possibility that the instrumentalisation of nature is not necessary to human flourishing, perhaps even more than some environmentalists would allow.

A plea for valuing nature is, of course, nothing new to contemporary environmentalists. However, the appreciation for the natural that is found in the modern consciousness of nature does not capture the significance of createdness that is at the heart of creation theology. The idea that creation is given to humanity as good, and that the human creature fits into the world, as the one responsible for it, goes far beyond any relationality or connectivity conceptualised in the natural sciences. An appreciation for nature as creation can be retrieved as a graced moment to be shared by all humanity, even in the face of current and even ongoing hardships and privation. Creation is to be savoured rather than manipulated; pragmatic considerations are always subordinate to that prior claim on our imagination. A faith context allows us to reach beyond the pragmatic (both our successes and our failures) to consider the assurance that the liberation of the human person is not just incidentally bound up with nature or the environment. Non-human creation is not external to the human person, and liberation is not freedom from nature but freedom with nature. Faith in the creator God relocates the human person in a cosmos, not of relentless matter, but one in which the free embodied person is ever called to be responsible for a gift already given.

Bibliography

Aldred, Jonathan. 'Existence Value, Moral Commitments and in-Kind Valuation'. In *Valuing Nature? Economics, Ethics and the Environment*, ed. John Foster. London: Routledge, 1997, pp. 155-69.
Alkire, Sabine. 'When Responsibilities Conflict: A Natural Law Analysis of Debt Forgiveness, Poverty Reduction, and Economic Stability'. *Studies in Christian Ethics* 14, no. 1 (2001): 65-80.
Anon. 'Stewardship'. In *The Oxford Dictionary of the Christian Church*, ed. E.A. Livingstone. Oxford: Oxford University Press, 1997.
Atkins, Margaret. 'Flawed Beauty and Wise Use: Conservation and the Christian Tradition'. *Studies in Christian Ethics; Ethics and Ecology* 7, no. 1 (1994): 1-16.
Avarez-Morales, A. 'Possible Implications of the Release of Transgenic Crops in Centres of Origin or Diversity'. *Environmental Biosafety Research* 2, no. 47-50 (2003).
Barrett, Scott. 'Montreal Versus Kyoto: International Cooperation and the Global Environment'. In *Global Public Goods: International Cooperation in the 21st Century*, ed. Inge Kaul, Isabelle Grunberg and Marc A. Stern. Oxford: Oxford University Press, 1999, pp. 192-219.
Barton, John. 'Virtue in the Bible'. *Studies in Christian Ethics* 12, no. 1 (1999): 12-22.
Bayer, Oswald. 'Self-Creation? On the Dignity of Human Beings'. *Modern Theology* 20, no. 2 (2004): 275-87.
Benzoni, Francisco. 'Thomas Aquinas and Environmental Ethics: A Reconstruction of Providence and Salvation'. *The Journal of Religion* 83, no. 3 (2005): 446-76.
Bloesch, Donald. 'A Review of Pannenberg's Systematic Theology, Volume 2'. *Christianity Today* 39, April (1995): 106.

Bonino, J.M. 'Natural Law'. In *Dictionary of the Ecumenical Movement*, ed. N. Lossky, J.M. Bonino, J. Pobee, T. Stransky, G. Wainwright and P. Webb. Geneva: World Council of Churches, 1991, pp. 712–15.

Bouchie, A. 'Organic Farmers Sue GMO Producers'. *Nature Biotechnology* 20 (2002): 210.

Bradley, Gerard. 'Moral Truth, the Common Good and Judicial Review'. In *Catholicism, Liberalism and Communitarianism*, ed. K. Grasso, G. Bradley and R. Hunt. Lanham, MD: Rowman and Littlefield, 1995, pp. 115–50.

Bretherton, Luke. *Hospitality as Holiness; Christian Witness Amid Moral Diversity*. Aldershot: Ashgate, 2006.

Brom, Frans W.A. 'Food, Consumer Concerns and Trust: Food Ethics for a Globalizing Market.' *Journal of Agricultural and Environmental Ethics* 12 (2000): 127–39.

Callicott, Baird. *In Defense of the Land Ethics*. Albany: State University of New York Press, 1989.

Catholic Bishops' Conference of England and Wales. 'The Common Good and the Catholic Church's Social Teaching'. London, 1996.

Cavanaugh, William. 'Killing for the Telephone Company: Why the Nation-State Is Not the Keeper of the Common Good'. *Modern Theology* 20, no. 2 (2004): 243–74.

Clifford, Anne. 'Anthropology: Contemporary Issues'. In *Handbook of Catholic Theology* ed. W. Beinert and F. Schüssler Fiorenza. New York: Crossroads, 1995, pp. 16–20.

—— 'Creation'. In *Systematic Theology: Roman Catholic Perspectives*, ed. Francis Schüssler Fiorenza and John P. Galvin. Dublin: Gill and Macmillan, 1992, pp. 195–248.

Colombo, Joseph. 'Systematic Theology'. *The Journal of Religion* 76, January (1996): 125.

Cooper, H., C. Spillane and T. Hodgkin, eds. *Broadening the Genetic Base of Crop Production*: CABI, 2001.

Day, Anil. 'Antibiotic Resistance Genes in Transgenic Plants: Their Origins, Undesirability and Technologies for Their Elimination from Genetically Modified Crops'. In *Transgenic Plants: Current*

Innovations and Future Trends, ed. C.N. Stewart. Wymondham: Horizon Scientific Press, 2003.

de Kathen, A. 'Biosafety Capacity Development and the Cartagena Protocol'. Paper presented at the Proceedings of the 6th International Symposium on the Biosafety of Genetically Modified Organisms, University of Saskatchewan, 2000.

De Schrijver, Georges. 'Combating Poverty through Development: A Mapping of Strategies'. Paper presented to School of Religions and Theology. Autumn, 2006.

——*Recent Theological Debates in Europe: Their Impact on Interreligious Dialogue*. Bangalore: Dharmaram Publications, 2004.

De Tavernier, Johan. 'Human or "Secular" History as a Medium for the History of Salvation or Its Opposite: "Outside the World There Is No Salvation"'. *Concilium: No Heaven without Earth* 4 (1991): 4–15.

Deane-Drummond, Celia, ed. *Aquinas, Wisdom Ethics and the New Genetics*. Edinburgh: T. & T. Clark, 2003.

——*Creation through Wisdom*. Edinburgh: T. & T. Clark, 2000.

——'Development and Environment: In Dialogue with Liberation Theology'. *New Blackfriars* 78, no. 916 (1997): 279–89.

——*Genetics and Christian Ethics*. Cambridge: Cambridge University Press, 2006.

——*The Ethics of Nature*. Oxford: Blackwell, 2004.

——*Theology and Biotechnology*. London: Cassell, 1997.

Deane-Drummond, Celia, Robin Grove-White, and Bronislaw Szerszynski. 'Genetically Modified Theology: The Religious Dimensions of Public Concerns About Agricultural Biotechnology'. *Studies in Christian Ethics* 14, no. 2 (2001): 23–41.

Deane-Drummond, Celia, and Bronislaw Szerszynski, eds. *Re-Ordering Nature; Theology, Society and the New Genetics*. Edinburgh: T. & T. Clark, 2003.

Department of Agriculture and Food. 'Co-Existence of GM and Non-GM Crops in Ireland.' Department of Agriculture and Food, http://www.agriculture.gov.ie/index.jsp?file=publicat/publications2005/gm_coexistence/index.xml.

Elshtain, Jean Bethke. 'Catholic Social Thought, the City and Liberal America'. In *Catholicism, Liberalism and Communitarianism*, ed. K. Grasso, G. Bradley and R. Hunt. Lanham, MD: Rowman and Littlefield, 1995, pp. 97–114.

Farrell, Edward P. 'Managing Our Forests for the Future'. Paper presented at the Annual Symposium of the Society of Irish Foresters, Newbridge, 1996.

Fergusson, David. 'Communitarianism and Liberalism: Towards a Convergence'. *Studies in Christian Ethics* 10 (1997): 32–48.

Fergusson, David S. *Community, Liberalism and Christian Ethics*. Cambridge: Cambridge University Press, 1998.

Finnis, John. 'Public Good: The Specifically Political Common Good in Aquinas'. In *Natural Law and Moral Inquiry: Ethics, Metaphysics and Politics in the Work of Germain Grisez*, ed. Robert P. George. Washington: Georgetown University Press, 1998, pp. 174–209.

Fletcher, G.L., S.V. Goddard, and C.L. Hew. 'Current Status of Transgenic Atlantic Salmon for Aquaculture'. Paper presented at the Proceedings of the 6th International Symposium on the Biosafety of Genetically Modified Organisms, University of Saskatchewan, 2000.

Forrest, Peter. 'How Kenotic Process Theology Underpins Humanist Deep Ecology'. *The Australasian Journal of Process Thought* 1, no. 1 (2000).

Foster, John, ed. *Valuing Nature? Economics, Ethics and the Environment*. London: Routledge, 1997.

Freyne, Sean. *Jesus, a Jewish Galilean: A New Reading of the Jesus Story*. Edinburgh: T. & T. Clark, 2004.

—— 'The Bible and Theology: An Unresolved Tension'. *Concilium: Unanswered Questions* 1, no. 1999/1 (1999): 15–20.

Fry, Timothy, OSB, ed. *The Rule of St Benedict*. New York: Vintage Books, 1998.

Ganoczy, Alexandre. 'Creation'. In *Handbook of Catholic Theology*, ed. Wolfgang Beinert and Francis Schüssler Fiorenza. New York: Crossroads, 1995, pp. 133–36.

—— 'Creation, Belief in, and Natural Science'. In *Handbook of Catholic Theology*, ed. Wolfgang Beinert and Francis Schüssler Fiorenza. New York: Crossroads, 1995, pp. 136–38.

—— 'Creation, Theology of'. In *Handbook of Catholic Theology*, ed. Wolfgang Beinert and Francis Schüssler Fiorenza. New York: Crossroads, 1995, pp. 144–46.

—— 'Ecology'. In *Handbook of Catholic Theology*, edited by W. Beinert and F. Schüssler Fiorenza, 201–3. New York: Crossroads, 1995, pp. 144–46.

Grasso, K., G. Bradley and R. Hunt, eds. *Catholicism, Liberalism and Communitarianism*. Lanhan, MD: Rowman and Littlefield, 1995.

Graumann, Sigrid. 'Experts in Bioethics and Biopolitics'. In *Bioethics in Cultural Context: Reflections on Methods and Finitude*, ed. Christoph Rehman-Sutter, Marcus Deuwell and Dietmar Mieth. Dordrecht: Springer, 2006.

Grove-White, Robin. 'The Environmental "Valuation" Controversy: Observations on its Recent History and Significance'. In *Valuing Nature? Ethics, Economics and Environment*, ed. John Forster. London: Routledge, 1997, pp. 21–31.

Gula, Richard. 'Natural Law Today'. In *Readings in Moral Theology Number 7*, ed. Charles Curran and Richard McCormick. New York: Paulist Press, 1991, pp. 369–89.

Halton, Thomas. 'Providence'. In *Encyclopedia of Early Christianity*, ed. Everett Ferguson. New York: Garland, 1999, pp. 957–58.

Hanemann, W.M. 'Valuing the Environment through Contingent Valuation'. *Journal of Economic Perspectives* 8, no. 4 (1994): 19–43.

Hardin, Garrett. 'The Tragedy of the Commons'. *Science* 162 (1968): 1243–48.

Harrison, Peter. 'Subduing the Earth: Genesis 1, Early Modern Science, and the Exploitation of Nature'. *Journal of Religion* 79 (1999): 86–109.

Hauerwas, Stanley. *Suffering Presence*. Edinburgh: T. & T. Clark, 1988.

Hiebert, Theodore. 'Re-Imaging Nature'. *Interpretation: A Journal of Bible and Theology* 50 (1996): 36–46.

Hilpert, Konrad. 'The Theological Critique of "Autonomy"'. *Concilium: The Ethics of Liberation – The Liberation of Ethics* 2, no. 172 (1984): 9–15.

Hodgson, G. 'Economics, Environmental Policy and the Transcendence of Utilitarianism'. In *Valuing Nature? Economics, Ethics and the Environment*, ed. John Foster, 48–63. London: Routledge, 1997.

Hoskins, W.G. *The Making of the English Landscape*. London: Hodder and Stoughton, 1955.

Houck, John W. and Oliver F. Williams, eds. *Co-Creation and Capitalism: John Paul II's Laborem Exercens*. London: University Press of America, 1983.

Hughes, G. 'Natural Law'. In *A New Dictionary of Christian Ethics*, ed. James F. Childress and J. MacQuarrie. London: SCM, 1987, pp. 412–14.

Hutchinson Edgar, D. 'Back to the Beginning: Paul's Eschatology in Romans'. Trinity College Dublin (unpublished manuscript), 2002.

Insole, Christopher. 'Against Radical Orthodoxy: The Dangers of Overcoming Political Liberalism'. *Modern Theology* 20, no. 2 (2004): 213–41.

Irish Council for Bioethics. 'Genetically Modified Crops and Food: Threat or Opportunity for Ireland?' 1–74. Dublin, 2005.

Isherwood, L. and E. Stuart. *Introducing Body Theology*. Sheffield: Academic Press, 1998.

Jacobs, Michael. 'Environment and Creative Value'. In *Valuing Nature? Economics, Ethics and Environment*, ed. John Foster. London: Routledge, 1997, pp. 232–46.

Jenczewski, E., J. Ronfort and A. Chevre. 'Crop-to-Wild Gene Flow, Introgression and Possible Effects'. *Environmental Biosafety Research* 2 (2003): 9–24.

Jia, S. and Y. Peng. 'Research and Regulation on Biosafety of GMOs in China'. *Environmental Biosafety Research* 2 (2003): 57–59.

Johnson, B. 'The Environmental Impact of GMOs: Some Possible Problems and Some Possible Solutions'. Paper presented at the Proceedings of the 6th International Symposium on the Biosafety of Genetically Modified Organisms, University of Saskatchewan 2000.

Johnson, Elizabeth. 'Powerful Icons and Missing Pieces'. In *Preserving the Creation: Environmental Theology and Ethics*, ed. K. Irwin and E. Pellegrino. Washington: Georgetown Press, 1995, pp. 60–66.
Junker-Kenny, Maureen. 'Christian Ethics and Culture'. In *The Critical Spirit; Theology at the Crossroads of Faith and Culture*, ed. Andrew Pierce and Geraldine Smyth, OP. Dublin: Columba Press, 2003, pp. 159–78.
—— 'Embryos *in Vitro*, Personhood and Rights'. In *Designing Life? Genetics in Human Reproduction*, ed. E. Graumann and S. Hildt. Aldershot: Ashgate, 1999, pp. 147–57.
—— 'The Image of God: Condition of the Image of the Human'. In *The Human Image of God*, ed. H.G. Ziebertz, F. Schweitzer, H. Häring and D. Browning. Leiden: Brill, 2001, pp. 73–82.
—— 'Valuing the Priceless: Christian Convictions in Public Debate as a Critical Resource and as "Delaying Veto" (J. Habermas)'. *Studies in Christian Ethics* 18, no. 1 (2005): 43–56.
—— 'Virtues and the God Who Makes Everything New'. In *A Wandering Galilean: Essays in Honour of Sean Freyne* (Leiden: Brill, 2009).
Kandawa-Schulz, M. 'Biosafety Regulations, Releases and Research Activities in Selected African Countries'. Paper presented at the Proceedings of the 6th International Symposium on the Biosafety of Genetically Modified Organisms, University of Saskatchewan, 2000.
Kaul, Inge, Isabelle Grunberg and Marc A. Stern. *Global Public Goods: International Cooperation in the 21st Century*. Oxford: Oxford University Press, 1999.
Keat, Russell. 'Values and Preferences in Neo-Classical Environmental Economics'. In *Valuing Nature? Economics, Ethics and the Environment*, ed. John Foster. London: Routledge, 1997, pp. 32–47.
Keeling, Michael. *The Mandate of Heaven: The Divine Command and the Natural Law*. Edinburgh: T. & T. Clark, 1995.
Keenan, R.J. and W.P. Stemmer. 'Nontransgenic Crops from Transgenic Plants'. *Nature Biotechnology* 20 (2002): 215–16.
Kerr, Fergus. *After Aquinas*. Oxford: Blackwell, 2002.

—— 'Moral Theology after Macintyre: Modern Ethics, Tragedy and Thomism'. *Studies in Christian Ethics* 8, no. 5 (1995): 33–44.
Keulartz, Josef. *Struggle for Nature: A Critique of Radical Ecology*. London: Routledge, 1998.
Kinderlerer, J. 'Transfer of Genetic Modification Technology into the Third World: Biosafety and Developing Countries'. Paper presented at the Proceedings of the 6th International Symposium on the Biosafety of Genetically Modified Organisms, University of Saskatchewan, 2000.
Kinderlerer, J. 'Why Regulate and How?'. *Environmental Biosafety Research* 2 (2003): 52–53.
Kioussi, J. 'The Evolution of the EU Regulatory System for GMOs.' Paper presented at the Proceedings of the 6th International Symposium on the Biosafety of Genetically Modified Organisms, University of Saskatchewan 2000.
Langemeyer, Georg. 'Creatureliness of the Human Person.' In *Handbook of Catholic Theology*, edited by Wolfgang Beinert and Francis Schüssler Fiorenza. New York: Crossroads, 1995, pp. 152–53.
—— 'Discipleship, Christian'. In *Handbook of Christian Theology*, ed. W. Beinert and F. Schüssler Fiorenza, 177–79. New York: Crossroads, 1995.
Levin, M.A. and H. Strauss. 'Overview of Risk Assessment'. In *Risk Assessment in Genetic Engineering*, ed. M.A. Levin and H. Strauss. New York: McGraw Hill, 1993, pp. 1–17.
Lisska, Anthony. *Aquinas's Theory of Natural Law: An Analytical Reconstruction*. Oxford: Clarendon Press, 1996.
Löser, Werner. 'Monasticism: Religious Life'. In *Handbook of Catholic Theology*, ed. W. Beinert and F. Schüssler Fiorenza. New York: Crossroads, 1995, pp. 487–89.
Lurquin, P. *The Green Phoenix; a History of Genetically Modified Plants*. New York: Columbia Press, 2001.
MacIntyre, Alasdair. 'A Partial Response to My Critics'. In *After Macintyre; Critical Perspectives on the Work of Alasdair Macintyre*, ed. J. Horton and S. Mendus. Notre Dame, IN: Notre Dame University Press, 1994.

MacNamara, Vincent. *Faith and Ethics: Recent Roman Catholicism*. Dublin: Gill and MacMillan, 1985.
—— 'The Distinctiveness of Christian Morality'. In *Christian Ethics*, ed. Bernard Hoose. London: Casells, 1998, pp. 149–60.
Matthews, Freya. *The Ecological Self*. London: Routledge, 1991.
McDonagh, Sean. *Passion for the Earth: The Christian Vocation to Promote Justice, Peace and the Integrity of Creation*. London: Geoffrey Chapman, 1994.
McFague, Sallie. *Super, Natural Christians; How We Should Love Nature*. London: SCM, 1997.
—— 'The World as God's Body'. *Concilium: Body and Religion* 2002, no. 2 (2002): 50–56.
Metz, Johann Baptist. *Faith in History and Society*. New York: Crossroads, 1980.
Mieth, Dietmar. 'Auf Dem Wege Zur Ethical Correctness: Vorbemerkungen Zur Ethik Des Libensmittelrechts Am Beispiel Der Novel Food-Problematik'. *ZLR Zeitschrift für das gesamte Liebensmittelrecht* 3 (1999): 267–86.
—— 'Autonomy of Ethics: Neutrality of the Gospel?'. *Concilium: Is Being Human a Criterion of Being Christian?* 5, no. 18 (1982): 32–39.
—— 'Autonomy or Liberation: Two Paradigms of Christian Ethics'. *Concilium: The Ethics of Liberation – The Liberation of Ethics* 2, no. 172 (1984): 87–93.
—— 'Bioethics, Biopolitics, Theology'. In *Designing Life? Genetics, Procreation and Ethics*, ed. Maureen Junker-Kenny. Aldershot: Ashgate, 1999, pp. 6–22.
—— 'Continuity and Change in Value-Orientations'. *Concilium: Changing Values and Virtues* 3, no. 191 (1987): 47–59.
—— 'Why We Cannot Allow the Cloning of Human Beings'. *Theologische Quartalschrift* 177 (1997): 1–6.
Milbank, John. 'The End of Dialogue'. In *Christian Uniqueness Reconsidered; The Myth of a Pluralist Theology of Religions*, ed. Gavin D'Costa. New York: Orbis, 1997, pp. 174–91.

Miles, Margaret. 'Body Theology'. In *A New Dictionary of Christian Theology*, ed. Alan Richardson and John Bowden, 77–78. London: SCM, 1983.

Millstone, E., E. Brunner, and S. Mayer. 'Beyond "Substantial Equivalence"'. *Nature* 401, October (1999): 525–26.

Mostert, Christiaan. *God and the Future; Wolfhart Pannenberg's Eschatological Doctrine of God*. London: T. & T. Clark, 2002.

Mueller, Joseph, SJ. 'Old Testament and Church Ministries in Two Ecumenical Dialogues'. *International Journal of Systematic Theology* 6, no. 3 (2004): 233–51.

Muir, W. and R. Howard. 'Assessment of Possible Ecological Risks and Hazards of Transgenic Fish with Implications for Other Sexually Reproducing Organisms'. *Transgenic Research* 11 (2002): 101–14.

Niebuhr, H. Richard. *Christ and Culture*. New York: Harper and Row, 1951.

Niemeyer, C.M. 'Semi-Synthetic Nucleic Acid–Protein Conjugate; Applications in Life Sciences and Nanotechnology'. *Reviews in Molecular Biotechnology* 82 (2001): 47–66.

Northcott, Michael S. '"Behold I Have Set the Land before You" (Deut. 1:8): Christian Ethics, GM Foods and the Culture of Modern Farming'. In *Re-Ordering Nature; Theology, Society and the New Genetics*, ed. Celia Deane-Drummond, Bronislaw Szerszynski and Robin Grove-White. Edinburgh: T. & T. Clark, 2003, pp. 85–106.

—— 'Concept Art, Clones, and Co-Creators; the Theology of Making'. *Modern Theology* 21, no. 2 (2005): 219–36.

—— 'Ecology and Christian Ethics'. In *Cambridge Companion to Ethics*, 209–27. Cambridge: Cambridge University Press, 2001.

—— *The Environment and Christian Ethics*. Cambridge: Cambridge University Press, 1996.

—— 'With the Grain of the Universe: The Church's Witness and Natural Theology by Stanley Hauerwas'. *The Expository Times* 114, no. 4 (2003): 129–30.

O'Donovan, Oliver. *Begotten or Made?* Oxford: Clarendon Press, 1984.

—— *Resurrection and Moral Order*. Leicester: Apollos, 1994.

—— *The Desire of the Nations*. Cambridge: Cambridge University Press, 1996.
O'Neill, Onora. *Autonomy and Trust in Bioethics*. Cambridge: Cambridge University Press, 2002.
—— 'Kantian Ethics'. In *A Companion to Christian Ethics*, ed. Peter Singer. Oxford: Blackwell, 1991, pp. 175–97.
Olson, R. 'Wolfhart Pannenberg's Doctrine of the Trinity'. *Scottish Journal of Theology* 43 (1990): 175–206.
Palmer, Clare. 'A Bibliographical Essay on Environmental Ethics'. *Studies in Christian Ethics* 7, no. 1 (1994): 68–97.
Pannenberg, Wolfhart. 'Problems between Science and Theology in the Course of Their Modern History'. *Zygon* 41, no. 1 (2006): 105–12.
—— *Systematic Theology Volume II*. Edinburgh: T. & T. Clark, 1994.
—— 'The Appropriation of the Philosophical Concept of God as a Dogmatic Problem of Early Christian Theology'. In *Basic Questions in Theology*. London: SCM Press, 1971, pp. 119–83.
—— *Towards a Theology of Nature*. Westminster: John Knox Press, 1993.
Paul, G. 'Communitarianism'. In *International Encyclopedia of Ethics*, ed. John K. Roth. London: FD, 1995, pp. 172–73.
Peterson, Gregory. 'The Created Co-Creator: What It Is and Is Not'. *Zygon* 39, no. 4 (2004).
Pollack, A. 'Plan for Use of Bioengineered Corn in Food Is Disputed'. *The New York Times*, 29 November 2000.
Porter, Jean. *Nature as Reason: A Thomistic Theory of the Natural Law*. Grand Rapids: Eerdmans, 2005.
—— 'Perennial and Timely Virtues: Practical Wisdom, Courage and Temperence'. *Concilium: Changing Values and Virtues* 3 (1987): 60–68.
—— *The Recovery of Virtue: The Relevance of Aquinas in Christian Ethics*. London: SPCK, 1994.
Queiruga, Andrés Torres, Lisa Sowle Cahill, Maria Clara Bingemer and Erik Borgman. 'What Is at Stake in Christology'. *Concilium: Jesus as Christ* 3 (2008): 7–10.

Quinn, P. *Divine Commands and Moral Requirements*. Oxford: Clarendon Press, 1978.
Radcliffe, S.J. 'A Professional Code of Ethics for the 21st Century'. *Journal of Forestry* July (2000): 16–21.
Rooney, Paul. 'Divine Commands, Natural Law and Aquinas'. *The Scottish Journal of Religious Studies* xvi, no. 2 (1995): 117–41.
Rosebergs, Alex. *A Philosophy of Science*. London: Routledge, 2000.
Russell, Cathriona. 'Review of Sean Mcdonagh, *Patenting Life? Stop! Is Corporate Greed Forcing Us to Eat Genetically Engineered Food?* Dublin: Dominican Publications, 2003'. *Administration* 52, no. 1 (2004): 108–10.
Schiemann, J. 'New Science for Enhanced Biosafety'. *Environmental Biosafety Research* 2 (2003): 37–41.
Schockenhoff, Eberhard. *Natural Law and Human Dignity; Universal Ethics in an Historical World*. Trans. Brian McNeil. Washington DC: Catholic University of America Press, 2003.
Sen, Amartya. *Development as Freedom*. Oxford: Oxford University Press, 1999.
——— 'Missing Women-Revisited'. *British Medical Journal* 327 (2003): 1297–98.
Siegfried, B.D. 'BT Transgenic Plants for Pest Management; Challenges and Opportunities'. Paper presented at the Proceedings of the 6th International Symposium on the Biosafety of Genetically Modified Organisms, University of Saskatchewan 2000.
Silva, P.S. and B. Buchanan. 'Regulatory Considerations for Transgenic Animal Health and Food Safety Assessment; a Canadian Perspective'. Paper presented at the Proceedings of the 6th International Symposium on the Biosafety of Genetically Modified Organisms, University of Saskatchewan 2000.
Smalla, K. 'Field Releases of Genetically Modified Micro-Organisms'. *Environmental Biosafety Research* 2 (2003): 65–68.
Smyth, S., G. Khachatourians and W. Phillips. 'Liabilities and Economics of Transgenic Crops'. *Nature Biotechnology* 20 (2002): 537–41.
Snow, A. 'Consequences of Gene-Flow'. *Environmental Biosafety Research*, no. 2 (2003): 43–46.

Sober, Elliot. 'Philosophical Problems for Environmentalism'. In *Environmental Ethics*, ed. Robert Elliot. Oxford: Oxford University Press, 1995, pp. 226–47.
Sowle Cahill, Lisa. *Between the Sexes*. Philadelphia: Fortress, 1985.
—— '"Embodiment" and Moral Critique'. In *Theological Issues in Bioethics*, ed. N. Messer. London: Darton, Longman and Todd, 2002, pp. 85–101.
—— 'Genetics, Individualism, and the Common Good'. In *Interdisziplinäre Ethik; Grundlagen, Methoden, Bereiche; Festgabe Für Dietmar Mieth Zum Sechzigsten Geburtstag*, ed. A. Holderreg and Jean-Pierre Wils. Freiberg: Herder, 2001, pp. 378–92.
—— 'Grisez on Sex and Gender'. In *The Revival of Natural Law*, ed. Nigel Biggar and Rufus Black. Aldershot: Ashgate, 2000, pp. 242–61.
—— 'On Richard McCormick: Reason and Faith in Post-Vatican II Catholic Ethics'. In *Theological Voices in Medical Ethics*, ed. Allen Verhey and Stephen Lammers. Michigan: Eerdmans, 1993, pp. 78–105.
Spash, C. 'Environmental Management without Environmental Valuation'. In *Valuing Nature? Economics, Ethics and the Environment*, ed. John Foster. London: Routledge, 1997, pp. 170–85.
Spinello, R. 'Deontological Ethics'. In *International Encyclopedia of Ethics*, ed. John K. Roth. London: FD, 1995, pp. 219–20.
—— 'Natural Law'. In *International Encyclopedia of Ethics*, ed. John K. Roth. London: FD, 1995, pp. 599–602.
Stewart, C. N., ed. *Transgenic Plants: Current Innovation and Future Trends*. Wymondham: Horizon Scientific Press, 2003.
Taylor, Charles. *Sources of the Self: The Making of Modern Identity*. Cambridge: Cambridge University Press, 1989.
Thompson, Paul. *Food Biotechnology in Ethical Perspective*. London: Blackie Academic and Professional, 1997.
—— *The Spirit of the Soil: Agriculture and Environmental Ethics*. London: Routledge, 1995.
Tomiuk, J. and A. Sentker. 'History of and Progress in Risk Assessment'. In *Transgenic Organisms: Biological and Social Implications*, ed. J.

Tomiuk, K. Wohrmann and A. Sentker. Basel: Birkhäuser Verlag, 1996, pp. 217–25.

Ullrich, Lothar. 'Resurrection of Jesus'. In *Handbook of Catholic Theology*, ed. W. Beinert and F. Schüssler Fiorenza. New York: Crossroads, 1995, pp. 591–95.

Vanin, Cristina. 'The Significance of the Incarnation for Ecological Theology: A Challenging Approach'. *Ecotheology* 6.1/6.2 (2001): 108–22.

Verspieren, Patrick. 'Dignity in Political and Bioethical Debates'. *Concilium: The Discourse of Human Dignity* 2 (2003): 13–22.

Verstraete, W. 'Environmental Biotechnology for Sustainability'. *Journal of Biotechnology* 94 (2002): 93–100.

Walsh, E.J. 'Genes, Plants and Mankind: A Plant Breeder's Perspective on the Reality and Potential of Plant Biotechnology'. In *Research Report of the Faculty of Agri-food and the Environment*. Dublin: University College Dublin, 2004, pp. 73–86.

Wheale, P. and R. McNally. *Genetic Engineering: Catastrophe of Utopia*. Hemel Hempstead: Harvester Wheatsheaf, 1988.

White, Lynn. 'The Historical Roots of Our Ecologic Crisis' Science 1967: 155 pp. 1203–1207.

Wolfe, Christopher. 'Subsidiarity: The "Other" Ground of Limited Government'. In *Catholicism, Liberalism and Communitarianism*, ed. K. Grasso, G. Bradley and R. Hunt. Lanham, MD: Rowman and Littlefield, 1995, pp. 81–96.

Worrell, Richard, and Michael C. Appleby. 'Stewardship of Natural Resources: Definition, Ethical and Practical Aspects'. *Journal of Agricultural and Environmental Ethics* 12 (2000): 263–77.

Wynne, Brian. 'Methodology and Institutions: Values as Seen from the Field'. In *Valuing Nature? Economics, Ethics and the Environment*, ed. John Foster. London: Routledge, 1997, pp. 135–52.

Zilinskas, R. and P. Balint. *Genetically Engineered Marine Organisms*. Boston: Kluwer Academic, 1998.

Index

Italics show where the topic is dealt with in detail.

Alkire, Sabine, 118, 119, 148
anthropocentrism, 13, 121, 142, 148,
  150, *161ff*, 205, 263, *see also*
  Environmental ethics
Aquinas, Thomas, 68, 96, 104, *106ff*, 120,
  232, 245, 258
Aristotelian ethics, 68, 87, 95, 97, 104ff,
  221, 258, 261
Atkins, Margaret, 182, *187ff*
autonomy in Christian ethics, *124ff*
  Christian distinctiveness in five
    dimensions, *128ff, 131ff, 264*
  commitment to a shared moral discourse, 136
  conductive method, *134ff*, 170
  content (or distinctive context), *129ff*
  definition and history, 124
  non-human creation from an autonomy perspective, 139
  obligation through freedom, *135ff*

Barth, Karl, *71ff*, 101
Barton, John, *259ff*
Bayer, Oswald, 266
Benzoni, Francisco, 121
biblical theology and the non-human
  creation, *152ff, see also* environmental ethics
biblical models in theology, 153
a new reading of Jesus' ministry for the
  liberation of all creation, *156ff*
Romans 8:19 and the 'salvation of all
  creation', *154ff*, 160

Bonhoeffer, Dietrich, 72
Bradley, Gerard, *112ff*

Christian communitarianism, *87ff*
  church as a source of Christian
    distinctiveness in ethics, *91ff*
  definition and history, *89ff*
  shared ecclesial ethical discourse, *98ff*
  teleological ethic, *95ff*
Christian context, 13, *65ff*, 78, 80
common good, 56, 62, 90, *114ff*, *246ff*
  sociability, *115ff*
  and subsidiarity, 114, *117ff*
creation, nature in theological perspective, *209ff, see also* Pannenberg
creation as a free act of the Trinitarian
  God, *214ff*
creatio ex nihilo, 215
God as Trinity, 216
God's preservation and rule of
  creation, 218
creation and wisdom, *211ff*
the God who 'makes everything new',
  15, 178, 211, 266
imago Dei, 191, 211, 212
the time-bound world of creatures,
  222
order and contingency, *191ff*, 212,
  223ff, 225, 231, *263ff*
the problem of evil in creation,
  228ff
the role of humanity in creation,
  226ff

creatureliness, *132ff, 226ff*
finite freedom, 63, 69, 83, 86, 137, 139, *265ff*

Deane-Drummond, Celia, 16, 69, 104, 145, 213, *250ff*, 270
   a biblical basis for a Christian virtue ethics, 259
   environmental theology based on wisdom, 255
   nature as creation in the 'virtue' ethics of, *251ff*
   stewardship reconstructed as the cultivation of the virtues, *254ff*
   theology as interpretative resource in conflicts on genetic technologies, 252
   a virtue ethic of nature as creation, 257
de Kathen, A., 49
De Schrijver, Georges, *93ff, 197ff, 206ff*, 268
De Tavernier, Johan, 132
development theory, 14
   and permanent economic growth, *197ff*
   sustainable, *192ff*
divine command ethics, *70ff, 129ff, 143ff*
   a specifically Christian discourse, *80ff*
   definition and history, *72ff*
   and non-human creation, *83ff*
   scripture as a source of Christian distinctiveness in ethics, *73ff*
   a theological deontology, *77ff*
dualism, 13, *171ff*
   Cartesian, 165
   Gnostic, 149
   and monism, 149, *172ff*

ecocentrism, 13, 150, *161ff, 205ff*, see also embodiment
   and body theology, *170ff*
   and radical ecology, *174ff*

embodiment, 12, 138, *170ff*, 245
environmental crisis, 121, *148ff, 163ff*
   'as a problem of modernity', *233ff*
   critiques of instrumentalism, 168
environmental ethics, *see also* biblical theology; stewardship; sustainability
   anthropocentrism in theological perspective, 167
   concepts and terms, 162
   critiques of instrumentalism, 168
   ecocentrism and body theology, 170
   environmental crisis, 163
   environmental or ecological ethics?, 166
   environmental theologies and reconstructions of stewardship, *149ff*
   models for environmental ethics, *161ff*
   theocentrism and environmental ethics, 177
environmental management, *see* strategies
Environmental Protection Agency (Ireland), 62

Fergusson, David, 93, 100
Finnis, John, 75, 120
Freyne, Sean, 13, *151ff*, 189
   a new reading of Jesus' ministry for the liberation of all creation, *156ff*
   biblical theology and the non-human creation, *152ff*

Ganoczy, Alexandre, 166
genetically modified food, *see* transgenic food organisms
Gula, Richard, *107ff*
Grasso, K. et al, 115

Index

Hardin, Garret, 193
Harrison, Peter, *185ff*
Hauerwas, Stanley, 92, 101
Hiebert, Theodore, 153
Hodgson, G., 55
Hutchinson-Edgar, David, *154ff*

Irish Council for Bioethics, 70, 203
Insole, Christopher, 99
institutional ethics, 69, 102, 118, 270
instrumentalism, 10, 13, 81, 168
  critiques of, *168ff*

Jacobs, Michael, 56
Junker-Kenny, Maureen, 69, 88, *96ff*, 138, 211, 268

Kant, Immanuel, *77ff*, *124ff*, 190, 268
  categorical imperative, *123ff*, *139ff*, 190
  'Kingdom of Ends', 11, 67, 128
Kaul, Inge, 175

Langemeyer, Georg, 189, 265
Lisska, Anthony, *95ff*, 112
Löser, Werner, 188
Lurquin, Paul, 22, 25

MacIntyre, Alasdair, 87, 89, *95ff*, 103
MacNamara, Vincent, 110, 131
  the question of content in Christian ethics, *129ff*
Magisterium, 105, 107, 110, 111
Malthus, Thomas, 193, 195
McDonagh, Sean, 183, 184
McFague, Sally, 172
Mieth, Dietmar, 69, *78ff*, *123ff*, *130ff*, 190, 231
  conductive method, *134ff*, 270
Milbank, John, *94ff*
Mostert, Christiaan, 265
Muir, W. and Howard. R., 39

natural law, *104ff*
  the common good and subsidiarity, 114, *116ff*
  definition and history, *106ff*
  deontology and teleology, *112ff*
  non-human creation and rational human nature, *119ff*
  scripture as a source in natural law ethics, *109ff*
  sociability, 115
  subsidiarity, *117ff*
nature, *see* creation
Northcott, Michael, 69, 87, 114, 121, 144, 176
  an evangelical reading of natural law ethics, 245
  the 'ecological crisis' as a problem of modernity, 233
  cultural encounter as a history of compromise, 235
  dominion, stewardship and Sabbath, 238
  resurrection as the vindication of creation, 240
  stewardship as the restoration of the order of creation, 238, *243ff*
  parochial ecology, *247ff*, 264

O'Donovan, Oliver, *71ff*, 116, *127ff*, *143ff*, *240ff*, 267
O'Neill, Onora, 140, 262

Pannenberg, Wolfhart, 15, *213ff*, 232, *see also* creation
Palmer, Claire, 150, 166
Porter, Jean, 69, *112ff*, 144, 178
public good, 18, *43ff*, *101ff*, 146, 198, 269

Rooney, Paul, 77

Schockenhoff, Eberhard, 246
Sen, Amartya, *194ff*, 207
   'Development as Freedom', 197, 268
   sustainability, population growth and food production, *193ff*
Sowle Cahill, Lisa, 69, 90, 105, 122, 129, 172
Spash, C., 54
stewardship, *149ff*
   from a Christian autonomy perspective, 190
   as domination or as co-creation?, 182
   as pilgrimage or service?, 187
   reconstructed as a cultivation of the virtues, 254
   as the restoration of the order of creation, 238
   and sustainability, 192
   towards a Christian anthropology of stewardship, 263ff
   and virtue ethics, 256
strategies in environmental management, *43ff*
   Cartagena Protocol 43, 48, 50
   economic strategies, 43
      consumer labelling, 20, 25, 42, 46, 49, 193, *199ff*
      contingent-valuation methods, *53ff*
      cost–benefit analysis 18, *43ff*, *51ff*, 62
      externalities 18, 34, *53ff*, 70
   regulation of transgenic organisms, *44ff*
      international perspectives on precaution and regulation, 45
      substantial equivalence in food safety, 45
      triggers for regulation, 46

sustainability, 58, 63, 182, 265
   as a 'concept of convergence', 10, 57, 63, 182
   development and permanent economic growth, 197
   intergenerational justice, 58
   normative aspects and descriptive aspects, *58ff*
   population growth and food production, *193ff*
   stewardship and, *192ff*

Taylor, Charles, 88, *90ff*, 103
theocentric ethics, 78, 121ff, 150, *161ff*, 177, 188, 205, 210, *248ff*, 264
theonomy, 12, 74, 86
theocentrism and environmental ethics, *177ff*, see also environmental ethics
Thompson, Paul, 47, 55, *57ff*
transgenic food organisms, *19ff*
   co-existence, *42ff*, *192ff*
   containment strategies, *25ff*
   Irish/EU context, 199
   techniques and applications, *21ff*
   transgenic modification of specific food organisms, 27
      fish and marine organisms, 38
      micro-organisms, 28
      plants, *30ff*
   risks and hazards, *18ff*, *27ff*, *38ff*
      foreseen but unintended consequences, *24ff*
      pleiotropic effects, *39ff*

virtue ethics, *102ff*, 113, *145ff*, *256ff*, see also Deane-Drummond, Celia

Walsh, Edward, *34ff*, 196
White, Lynn, *181ff*